The Stars and Serendipity

THE STARS & SERENDIPITY

by Robert S. Richardson
Associate Director, Griffith Observatory, Los Angeles, California

Illustrated with photographs and drawings

Pantheon Books

ACKNOWLEDGMENTS

Figure 1: courtesy of the Lewis Walpole Library and Commander C. Campbell-Johnston. Figure 13: Lick Observatory photograph. Figure 23: courtesy of Cave Optical Company. Figure 30: Sproul Observatory photograph by Sarah L. Lippincott and John Wooley. Figures 34, 35, 42, 43, 44, 45, 46, 48, 49a and b, 51: photographs from the Hale Observatories. Figure 37: courtesy of American Museum of Natural History. Figures 38, 39, 40a-d, 41: courtesy of Jet Propulsion Laboratory, California Institute of Technology. Figure 50: courtesy of Grote Reber. Figure 52: courtesy of Science, *vol. 164, 9 May 1969, copyright 1969 by the American Association for the Advancement of Science. Jacket photograph from the Hale Observatories, copyright by the California Institute of Technology and Carnegie Institution of Washington.*

COPYRIGHT © 1971 BY ROBERT S. RICHARDSON

*All rights reserved under International and Pan-American Coypright Conventions.
Published in the United States by Pantheon Books, a division of Random House, Inc., New York
and simultaneously in Canada, by Random House of Canada Limited, Toronto.*

Trade Ed.: ISBN: 0-394-82022-3   Lib. Ed.: ISBN: 0-394-92022-3

Library of Congress Catalog Card Number: 70-77435

Manufactured in the United States of America

Designed by Janet Townsend.

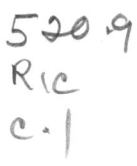

Contents

Introduction: Discoveries that Didn't Make Sense		3
1.	Sir Horace and Serendipity	5
2.	The Not-so-Eternal Stars	9
3.	c = Velocity of Light	30
4.	Uranus and Beyond	44
5.	Sunspots	55
6.	The Strange Companion of the Dog Star	60
7.	Those Canals of Mars	81
8.	The Solar Magnetic Cycle	92
9.	The Shift Is to the Red	99
10.	Galactic Radio Waves	108
11.	Radio Bursts from Jupiter	111
12.	Fireball Radiation	113
13.	Pulsars	115
14.	"Mysterium"	118
15.	Conclusion	122
	Glossary	125

The Stars and Serendipity

Introduction: Discoveries that Didn't Make Sense

You can buy all kinds of books on science now. Books that tell you why a man weighs less on the moon than on the earth, why a whale is not a fish, and how to tell temperature by counting the chirps of a cricket. When there are so many popular-science books on the market why write another?

Well . . . we believe this is a different kind of popular-science book. Most such books make the "Whys" and "Hows" of science sound too easy. As if the answers to everything were already known. As if the results of science always agreed with prediction. Almost as if a scientist knew what he was going to discover before he discovered it.

Sometimes science does work that way. The planet Neptune was discovered by prediction from Isaac Newton's theory of gravitation. A mathematician told astronomers where to look for Neptune among the stars. The astronomers looked and there was Neptune—just as predicted.

But just as often—perhaps even more often—the great discoveries of science came about in an entirely different way. Far from being discovered as the result of prediction from theory, their discovery was wholly unpredicted and came as a complete surprise. A scientist started an investigation hoping to find one thing and instead found something else entirely different. Entirely different because nobody with any sense would have predicted that such a thing existed in the first place. For, contrary to what most people think, science does not always move ahead in a "sensible" way.

In this book we are going to talk about some discoveries that scientists made, not by following theory but by pursuing slight devi-

The Stars and Serendipity

ations from accepted theory, whether they seemed to make sense or not. We shall talk about discoveries in astronomy because we are most familiar with astronomy. But our remarks could apply equally well to discoveries in medicine, chemistry, botany, or any other field of research you might name. Even to inventing better ways of doing housework, for instance.

The name for this happy business of making unanticipated discoveries is—of course—*serendipity*.

1 · Sir Horace and Serendipity

How most words got started is a mystery. But there is no mystery about serendipity. We can date its beginning to the very day. The word first occurs in a letter written by Sir Horace Walpole on Monday, January 28, 1754.

First a few words about Sir Horace himself.

He was born in 1717, the youngest of the five children of the fourth Earl of Orford, an elegant little boy who calmly accepted the fashionable aristocratic world in which he was reared as his natural birthright. He was also fully aware of its advantages. Although of delicate health and chronically troubled by gout, he lived until 1797, so that his life spanned nearly the entire eighteenth century. (Someone has said that the best way to attain old age is to be born with a minor chronic ailment and then coddle it.)

Walpole was quite different from our usual picture of the young eighteenth-century British aristocrat. He had no desire to fritter away his estate in idle dissipation. Good heavens, no! He had very different ideas.

Except that he never married, Sir Horace lived a kind of life I would rather enjoy myself. The country gentleman sitting in his luxurious library (Fig. 1), writing and reading whatever interested him, free from telephone calls, the neighbor's radio, and other distractions. He wrote continuously, and unlike most writers, never had any worries about getting his work published. He owned his own printing press.

Walpole's best-known work is his novel *The Castle of Otranto*, which he claimed to have written in a kind of trance-like condition

in two months. It is a fantastic tale of the goings-on in an ancient stone castle complete with all the weird effects you might expect to find in an ancient stone castle. A sighing portrait that steps from its canvas, an armored hand that comes creeping up the staircase banisters, a gigantic black-feathered helmet that crashes down in the castle courtyard, etc. Wandering through the gloomy halls of Otranto we meet the evil Prince Manfred, some beautiful maidens, a mysterious stranger, and other tormented souls in anguish.

Sir Horace published *The Castle of Otranto* anonymously in 1764, but it proved such a success that he got out a second edition the following year under his own name. Although the public liked it, critics have been unanimous in pointing to *The Castle of Otranto* as an outstanding example of the Gothic novel at its worst.

Walpole is remembered today, however, not for his novels and essays, but for his letters. Most people I know hate to write letters. Sir Horace loved to write letters. He must have tossed off hundreds, maybe thousands, of letters, filled with gossip about people of note. Although few of these people are of any importance today, his comments on them furnish us with a valuable source of information on life among the British aristocracy of the eighteenth century.

As a matter of historic interest we quote here the first, or "serendipity," part of a letter that Walpole wrote to Sir Horace Mann on the occasion of the arrival of an old portrait of Bianca Capello, for which Walpole had ordered a special frame. Bianca Capello (1548–87) was the beautiful wife of Francesco de Medici, Grand Duke of Tuscany, and Walpole thought it appropriate to have the coat of arms of the two families portrayed on the two sides of the frame.

While searching in an old book of Venetian arms he discovered that

... there are two coats of Capello, who from their *name* bear a *hat*, on one of them is added a flower-de-luce on a blue ball, which I am persuaded was given to the family by the Great Duke, in consideration of

Fig. 1. Horace Walpole in the library of his "little Gothick castle," of Strawberry Hill on the Thames, which he bought in 1749. It is now a Catholic girls' school.

The Stars and Serendipity

this alliance; the Medicis you know bore such a badge at the top of their own arms . . . This discovery indeed is almost of that kind which I call *serendipity*, a very expressive word, which as I have nothing better to tell you, I shall endeavor to explain to you: you will understand it better by the derivation than by the definition. I once read a silly fairy tale, called *The Three Princes of Serendip:* as their hignesses travelled, they were always making discoveries, by accidents and sagacity, of things which they were not in quest of: for instance, one of them discovered that a mule blind of the right eye had travelled the same road lately, because the grass was eaten only on the left side, where it was worse than on the right—now do you understand *serendipity?* One of the most remarkable instances of this *accidental sagacity* (for you must observe that *no* discovery of a thing you *are* looking for, comes under this description) was of my Lord Shaftsbury, who happening to dine at Lord Chancellor Clarendon's, found out the marriage of the Duke of York and Mrs. Hyde, by the respect with which her mother treated her at table. I will send you the inscription in my next letter; you see I endeavor to grace your present as it deserves.

And so on and on for much more gossip.

Thus was serendipity born, a word tossed off by Sir Horace Walpole in a carefree moment without a thought of its possible survival.

2 · The Not-so-Eternal Stars

"Since finding the year's time length is the first of all the things demonstrated concerning the sun, we shall first learn from the treatises of the ancients the disagreements and difficulties concerning their statements on this, and especially from that of Hipparchus, a diligent and truth-loving man."

Ptolemy, The Almagest, Book Three, c. A.D. 150

Most people can point out a few constellations, the Big Dipper and Little Dipper, the W-shaped Cassiopeia, the belt and sword of the giant Orion, and the little cluster called the Pleiades or Seven Sisters. Another familiar one is the winding line of stars that form the Scorpion, its heart marked by the red Antares. Don't wonder if there is something wrong with your eyes if you can't see these figures the ancient people put in the sky. Astronomers can't see them either. In fact, there would be something wrong with your eyes if you *could* see them. The constellations are so old that nobody knows how they got their names, but certainly it was not because of their fancied resemblance to some person or object. Mount McKinley doesn't bear the least resemblance to our twenty-fifth President. Neither does Reindeer Lake in Canada resemble a reindeer.

You always associate the constellations with the seasons. Winter

Fig. 2. Overleaf: The appearance of stars looking north on May 10, 1970, about nine o'clock in the evening.

URSA MAJOR
URSA MINOR
North Celestial Pole
Polaris
CASSIOPEIA
WEST

Lyra
Vega

Deneb
CYGNUS

Mt. San Antonio

EAST →

The Stars and Serendipity

is the time of year when the stars of Orion sparkle so brightly on clear frosty nights. The Scorpion belongs to the warm evenings of summer. Halloween is when the Pleiades are riding high. Each year the same stars return to the same places at the same times you saw them there the year before. No wonder we speak of the "eternal stars."

Eleven Thousand Years in the Future

You stroll out in your backyard for a look at the stars before going to bed. The time is nine o'clock of May 10. The place is somewhere in southern California. Looking north you see the stars as shown in Figure 2. Polaris, the North Star, is straight ahead about one-third the distance from the horizon to the zenith. Polaris is at the end of the curving line of four stars that form the handle of the Little Dipper. As the earth turns, the Little Dipper appears to swing around the heavens in a counterclockwise direction as if fastened to the sky at the end of its handle by Polaris. Except for Polaris and the two stars in the end of the bowl, the stars of the Little Dipper are rather faint and hard to locate.

But you can't miss the seven stars that outline the Big Dipper, which is now hanging bottom-up over Polaris. Below Polaris and opposite the Big Dipper we see some of the stars of Cassiopeia, although most of the "W" is hidden by the San Gabriel Mountains. On your right, toward the northeast, the bright blue star Vega is flashing over Mount San Antonio. And to think! A few years ago, before you joined the Azusa Astronomical Society, you couldn't even identify Polaris.

The air is so balmy and you feel so relaxed that you decide to stretch out on the grass for a while. But you mustn't stay out long. You've got that early appointment . . .

You waken suddenly with a vague sense that something is wrong. It was dark when you went to sleep. Now there is sufficient light to see all around you. Have you slept all night? But, no, the twilight is fading instead of brightening.

The Not-so-Eternal Stars

But where are you? Gone is your backyard . . . gone is everything familiar. You are in an utterly strange world. Occasionally a figure passes by, garbed as if for a masquerade. Can these be the creatures that dwell in those curiously misshapen huts over there? With mounting alarm the conviction grows that you have been asleep for a length of time compared with which Rip Van Winkle's snooze of twenty years was a mere catnap.

But isn't that mountaintop on your right Mount San Antonio? And the horizon line is remarkably similar to the San Gabriel range. Maybe some of these queer-looking people would know. But when you question them they shake their heads, mutter something unintelligible, and hurry on.

It's night now. The air is warm and heavy with the scent of trees in blossom. Must be the spring of the year. But *what* year? You'll never know. It's hopeless.

Is it hopeless? You remember a lecture at the Azusa Astronomical Society. It's possible to make a guess, now that the stars are out. There they are—the Big Dipper and Cassiopeia and Vega and Polaris at the end of the Little Dipper. How good they look! Still exactly the same.

But although the constellations *look* the same, after a few hours you find they aren't *behaving* the same (Fig. 3). True, the Big Dipper and Cassiopeia are on opposite sides of Polaris as they should be. Only they aren't revolving around Polaris as a center. Neither is the Little Dipper turning around Polaris. The star at the end of the handle of the Little Dipper has been displaced from its central position in the northern heavens. Toward morning there can't be any doubt. The earth has a magnificent new Polaris—the brilliant blue Vega.

Your sleep was a long one indeed—about 11,000 years long. For if Vega has become the North Star, you are in the world of A.D. 13,500.

Fig. 3. Overleaf: The appearance of stars looking north on May 10, 13,500.

Dene
CYGNUS

Aquila Altair

Lyra Vega
North Celestial Pole

HERCULES

CASSIOPEIA

URSA MINOR

URSA MAJOR

Mt. San Antonio

The Stars and Serendipity

A Clock Called Precession

How could an amateur know enough about astronomy to determine the date so far in the future? From naked-eye observations, too.

The answer is "precession." You can think of precession as a clock that takes 26,000 years to go around the dial. From a knowledge of precession you can estimate the date fairly accurately, within a thousand years anyhow.

Precession is short for "precession of the equinoxes," an apparent motion of the celestial sphere discovered by Hipparchus of Nicaea about 127 B.C. Hipparchus, by common consent, is called the Father of Astronomy and the precession of the equinoxes regarded as his greatest discovery.

Apparently Hipparchus became aware of precession in the course of another investigation. We are on admittedly shaky ground here for we have no way of knowing just what Hipparchus' mental processes were on this matter. None of his work has survived. All we know about Hipparchus is what others who came a century or two later have told us. It seems that he was engaged in trying to improve the calendar, and to do so he needed a more accurate value for the length of the year. (Incidentally, astronomers are still trying to improve the calendar.) Perhaps you thought there is just "the year." Actually there are half a dozen different kinds of year. There are only two that concern us here: the sidereal year, or the year measured by the stars, and the year of the seasons. The length of both years is found from observations of the sun.

Let us take first the year by the stars, or the sidereal year.

Suppose the stars were visible in the daytime, as they would be were it not for our atmosphere. Out in space or on the airless surface of the moon we know the stars are always visible even when right around the sun. The date is August 20. Looking at the sun we see it appears to be almost touching the bright star Regulus in the constellation of Leo (Fig. 4). The next day the sun has moved twice its diameter to the east of Regulus, and the day following is four diameters east of the star. The sun continues eastward until, by

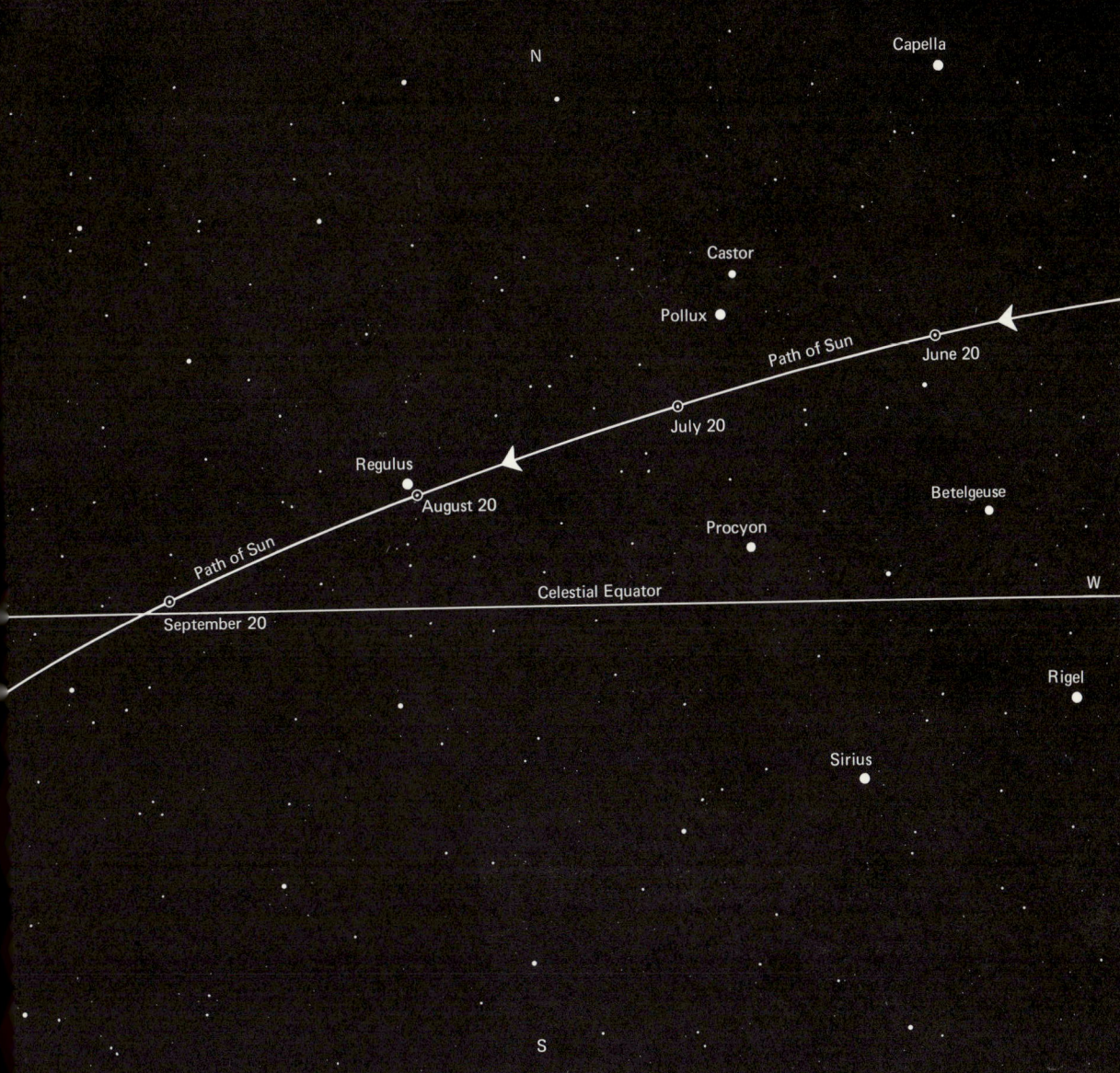

Fig. 4. The sun would be near bright star Regulus on August 20, if stars were visible in daytime. (As they would be were it not for the earth's atmosphere.) Sun is apparently moving east about twice its diameter per day.

October 15, it is far from Regulus and near the bright star Spica in Virgo, and by November 30 we find the sun near the bright red Antares in Scorpio. By August 20 the sun is back near Regulus again.

The sun has now apparently made a complete circuit of the

The Stars and Serendipity

heavens with respect to the star Regulus. The time required was 365 days 6 hours 9 minutes 9.5 seconds. Astronomers call this time interval the sidereal, or star-to-star, year. (Any other star would have done as well as Regulus. It just happens that the sun passes very close to Regulus.) The sidereal year is the true period of time required for the earth to revolve around the sun.

Another kind of year, and the one that is most important to us, is the year of the seasons. Often about March 20 you see a little paragraph in the paper stating that "according to astronomers at the Mt. Washington Observatory spring officially begins here tomorrow at 3:17 P M." How do these astronomers know when spring "officially" begins? Today it is easy. They just look it up in a book published each year by the U.S. Naval Observatory. But the old astronomers had no such source of information. They had to determine the beginning of spring (or summer, fall, or winter) by measuring the length of the shadow cast by the "gnomon" (Fig. 5). The gnomon is believed to be the world's oldest scientific instrument. It certainly was not a very sophisticated type of instrument. In fact, it was simply a tall stake set in a level space of ground. Yet it is amazing how much information you can get out of such a simple device.

Fig. 5. Time of day and beginning of seasons could be determined from the shadow cast by a gnomon.

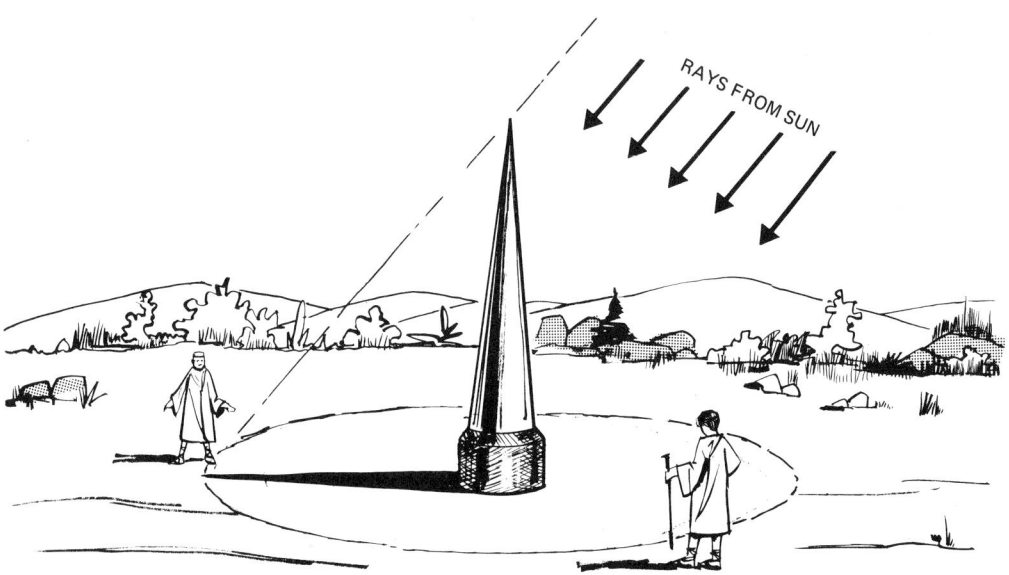

The Not-so-Eternal Stars

Suppose the earth's equator were painted on the surface so that you could follow its course, like the white line that marks the center of the highway. Imagine this line expanding upward and outward from the surface until it cuts the sky in the celestial equator (Fig. 6). If you were standing on the equator of the earth you would see the celestial equator directly overhead arching out of the horizon in the east and west.

If the earth revolved around the sun with its axis vertical or straight-up-and-down relative to its orbit, like a soldier marching stiffly upright around the parade ground, then no matter what the time of year the sun would always be somewhere on the celestial equator. The days and nights would all be equal. There would be no seasons. And the earth would be a pretty dull place to live.

But somehow, long ago, the earth got its axis of rotation tilted over by 23½ degrees from the vertical to its orbit. In June the north end of the axis is tilted toward the sun. Then for people like ourselves in the northern hemisphere, at noon the sun is almost overhead, and its rays beat nearly straight down on the surface. Daylight lasts for fifteen hours. But six months later, in December,

Fig. 6. Imagine the equator of the earth expanding outward until it cuts the celestial sphere of the stars. This imaginary circle in the heavens is the celestial equator.

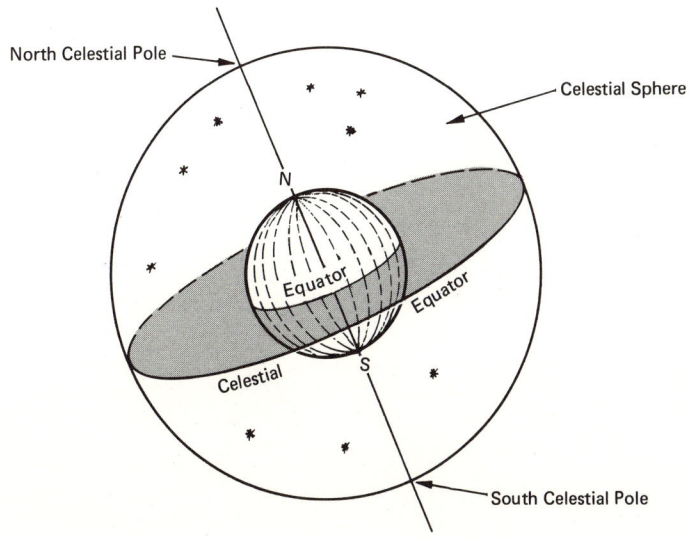

The Stars and Serendipity

when the earth is on the opposite side of its orbit, the north end of its axis is tilted away from the sun. At noon the sun is low in the sky and its rays strike the surface at a slant. Daylight lasts for only nine hours. Hence in the United States it is colder in December than in June. Of course, the opposite conditions prevail south of the equator.

Suppose the sun traced out a visible line in the sky so that we could follow its eastward motion from day to day among the stars. We start tracking the sun late in December when it is farthest south of the celestial equator. We soon find the sun is not only moving east but also toward the north. The sun approaches the celestial equator, slowly at first then faster, crossing it from south to north usually on March 21 (Fig. 7). Spring begins at the moment of crossing. The point of crossing, or vernal equinox, is marked by the sign γ, representing the horns of a ram.

Fig. 7. The sun in eastward motion crosses the vernal equinox about March 21, when spring begins in northern hemisphere. Although the vernal equinox is now in the constellation of Pisces, we still use the symbol γ, the horns of a ram, from the time when the vernal equinox was in Aries the Ram.

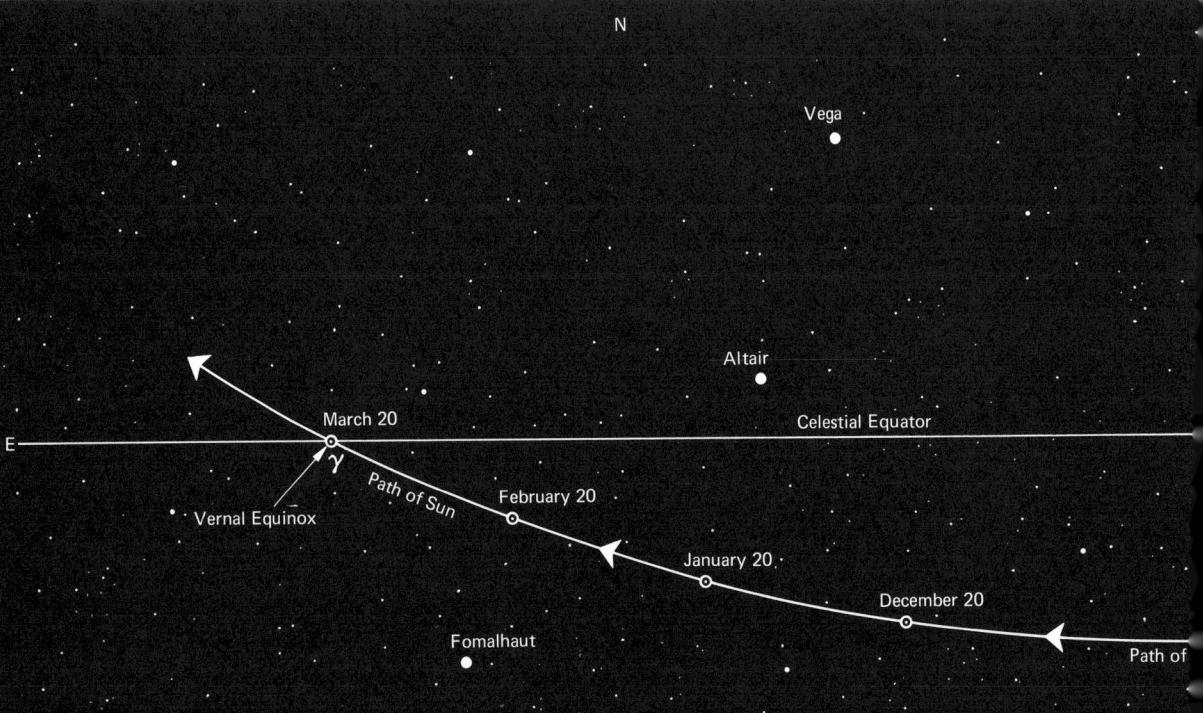

The Not-so-Eternal Stars

In the course of his work on the calendar, Hipparchus made a discovery that must have come as quite a surprise to him. One that he could hardly have anticipated when he started out. What Hipparchus discovered is that the sun does not cross the celestial equator at the same point each year. That is, the position of the vernal equinox is not fixed but is moving steadily westward. If this year the sun crossed the celestial equator exactly in line with some star, next year the point of crossing would be slightly west of the star (Figs. 8 and 9). It is as if the vernal equinox moved back a little each year to meet the sun. Thus the year from star to star is a little longer (twenty minutes) than the year from spring to spring, or the year of the seasons. The length of the year of the seasons is 365 days 5 hours 48 minutes 46 seconds.

Fig. 8. Below left: An observer sees the sun cross the celestial equator from south to north on March 21, at beginning of spring. Vernal equinox is at γ_1
Fig. 9. Below right: A year later the observer sees the sun cross the celestial equator slightly west of its previous position on March 21. Vernal equinox has shifted westward to γ_2. (Westward motion of the vernal equinox is enormously exaggerated.)

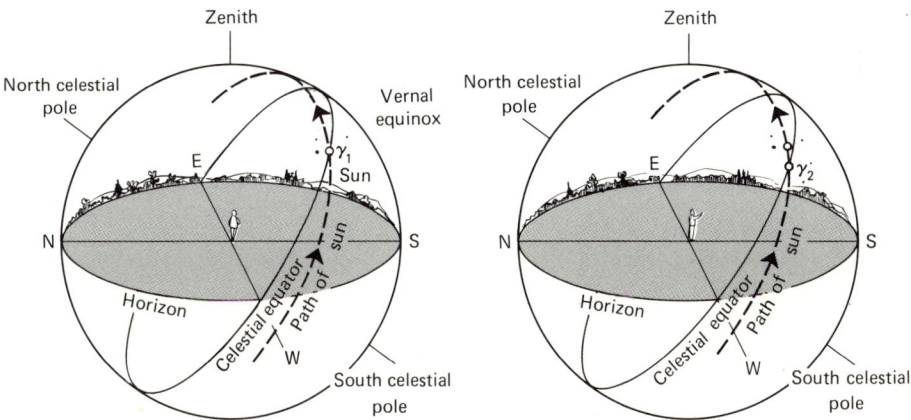

The Stars and Serendipity

Hipparchus was led to this discovery by comparing his observations of the sun and stars with those of Timocharis of Alexandria made 150 years earlier. (Hipparchus probably regarded Timocharis as an "old" astronomer.) The extra distance the sun has to move each year from the vernal equinox to a star of twenty minutes is so small that it would pass unnoticed from rough observations made a few years apart. But in the 135 years between the observations of Timocharis and those of Hipparchus the westward shift of the vernal equinox had built up to four times the apparent width of the full moon, too large an amount to be overlooked.

This westward motion, or "precession of the equinoxes," as it came to be called, has kept astronomers busy ever since Hipparchus announced its discovery some two thousand years ago. For a long time their main job was trying to fix the *amount* of the shift as accurately as possible—*how much* the equinox moves westward per year. An explanation of *why* the vernal equinox moves westward as it does was not forthcoming until Isaac Newton published his *Principia* on May 8, 1686. (*Principia* is short for the *Mathematical Principles of Natural Philosophy*.)

The Many Motions of the Earth

What is the cause of precession? Why does the vernal equinox keep shifting westward?

The complete theory of the earth's precessional motion is extremely complex. We won't try to explain to you why the axis of the earth moves as it does. But perhaps we can make it seem in some degree reasonable that it should.

Back in the old vaudeville days there was a magician whose trick suitcase was a never-failing source of publicity for his act. Alighting from a train he would casually hand the suitcase to a redcap. The suitcase was not particularly heavy or unusual-looking in any way. Yet the redcap found it completely unmanageable. If he tried to turn a corner the suitcase would not allow itself to be turned. If he tried to throw it on a baggage truck it would nearly twist out of his hand.

The Not-so-Eternal Stars

The suitcase—entirely obedient for the magician—seemed possessed of a demon at the redcap's touch.

The suitcase contained a motor-driven gyroscope controlled by a secret switch. A gyroscope consists of a heavy wheel mounted in such a way that its axis of rotation can be set in any desired direction (Fig. 10). If the wheel is motionless, the gyroscope can be moved as readily as any other object of the same weight. But if spinning rapidly, the gyroscope stubbornly resists any turning force applied to it. If free to turn, the axis of the wheel will maintain its original direction in space, even though the mounting of the gyro as a whole is changed in position.

To make the wheel change position we must apply a turning force to its axis of spin (Fig. 11). The behavior of the wheel is quite different from what you would expect. You would *expect* the wheel to start tipping under the pull of the weight. Instead the axis of spin moves off in a direction *at right angles* to the direction of pull, quite surprising when you see it for the first time. Reversing the spin of the wheel causes its axis to move in the opposite direction.

Fig. 10. Below left: If the wheel of a properly mounted gyroscope is set spinning with axis of rotation pointed in any direction, the supporting frame can be tilted in any way desired, but whirling axis will keep on pointing in original position.

Fig. 11. Below right: Contrary to expectation the downward pull of gravity, instead of causing gyroscope to topple over, sets it in motion in a horizontal circle.

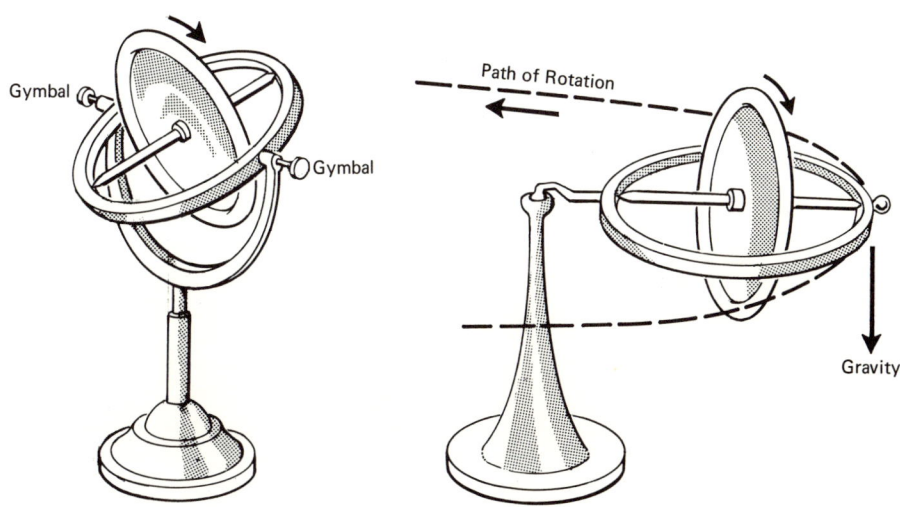

The Stars and Serendipity

You will have to take our word for it that two rotations acting at right angles on a body—the spin of the wheel and the pull of the weight—combine to produce a third rotation at right angles to them both. This explains why a spinning top does not fall over under the pull of its own weight. Instead of falling over, the top's axis of spin begins describing a circular motion having the shape of an ice-cream cone. But a physicist would not describe it that way. He would say the top "precesses."

The Moon's Pull

Astronomers have found that the axis of rotation of the earth is also describing a conical motion in space, or precessing. For the earth to precess there must be two rotations acting upon it at right angles. One of these is plainly the daily rotation of the earth upon its axis. But what is the other? What other force is acting upon the earth corresponding to the downward pull of gravity on the spinning top?

This other force is found in the gravitational pull of the moon.

If the earth were a uniform sphere from center to surface like a solid rubber ball, the attraction of the moon would be equivalent to a single force passing through its center, and there would be no precession.

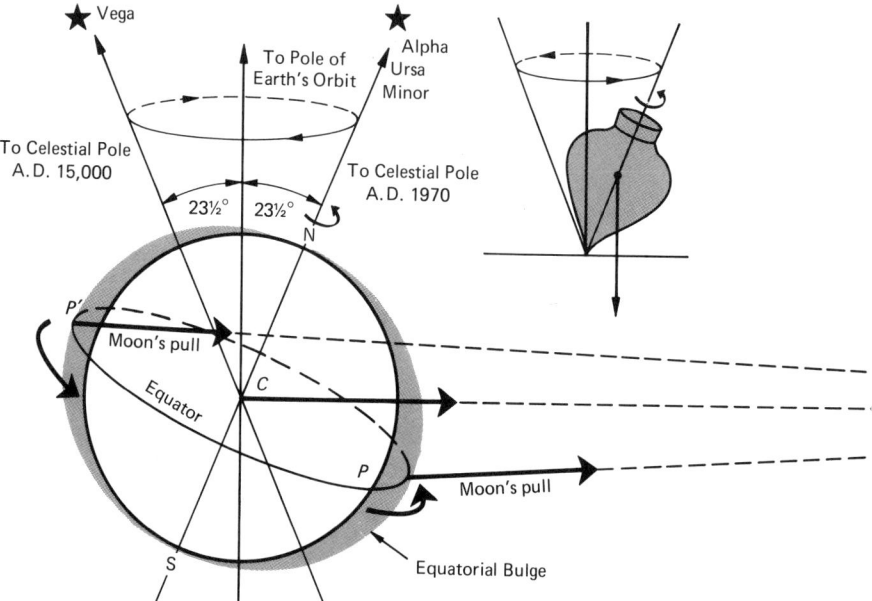

But the earth is not a sphere. The earth is a body with somewhat the shape of a grapefruit, slightly flattened at the poles and bulged around the equator. Such a body is called a *spheroid*. Viewed in a telescope it is easy to see that the planet Jupiter is a spheroid. But the earth's equatorial bulge is so slight you would never be able to detect it simply by casual inspection. Yet this bulge, slight as it is, makes a big difference in the effect of the moon's attraction on the earth.

Figure 12 shows the earth with its equatorial bulge vastly exaggerated. The moon attracts the particle in the bulge at P harder than the more distant particle at C. For the same reason, the moon attracts the particle C at the center harder than the particle in the bulge at P'. Assume for the moment that the earth has no rotation about its axis NS. What would happen? Since the moon is pulling harder on P than on P', the result of the moon's pull would be to "straighten up" the earth, so as to make its axis NS perpendicular to its orbit. That is, to give the earth a rotation around an axis through C.

But the earth *is* rotating . . .

The result of these two rotations acting at right angles is to make the axis NS move off at right angles to the direction of the moon's pull. The rotational effect of the moon is very weak and hence the

Fig. 12. The rotation of the earth, combined with the attraction of the moon and sun on the earth's equatorial bulge, causes the axis of the earth to change direction in space without altering its tilt of axis to orbit. The slow, turning motion, called "precession," completes a revolution in about 26,000 years. A spinning top also shows precession.

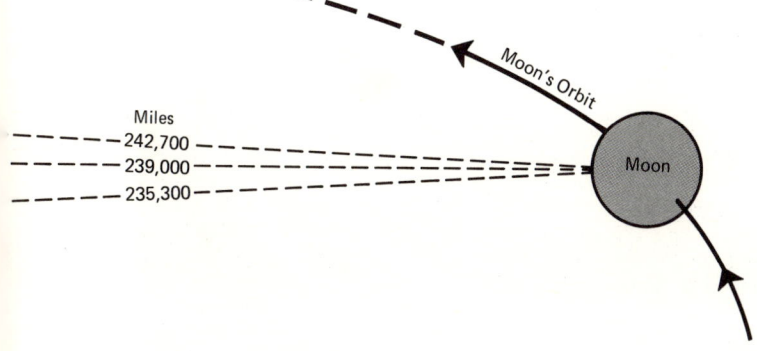

The Stars and Serendipity

precessional motion is correspondingly slow. So slow that the earth's axis prolonged requires about 26,000 years to complete a revolution in space.

Precession is so slow that we can only detect its effect in changing star positions over an interval of a century or more from naked-eye observations. With photography the precessional motion is apparent from observations only a few years apart. Figure 13 shows the circumpolar star trails recorded in exposures of nine hours taken with a fixed camera directed at the north celestial pole. Our North Star, Polaris, made the bright streak near the center of the picture. This picture shows two exposures, one taken in 1907, the other in 1941, superimposed on each other. You can see clearly the significant differences in the trails on the two exposures.

Fig. 13. The North Pole approaches Polaris. Star trails photographed at the Lick Observatory by J. C. Duncan, 1907, and J. F. Chappell, 1941. Photographs superimposed so that the beginnings of the two trails of each star coincide.

The Not-so-Eternal Stars

Right now precession is causing the north celestial pole to approach Polaris until by A.D. 2102 (Fig. 14) their distance will be minimal. Their separation will then be reduced to about the apparent width of the full moon. The north celestial pole will now begin to move away from Polaris and after a thousand years it will be so far from this star we can hardly call it "Polaris" any longer. In fact, there will be intervals stretching over thousands of years when the north celestial pole will not be near any bright star and the earth will have no Polaris. We will have the most splendid Polaris about A.D. 13,900, when the direction of the earth's north axis will point nearest Vega, a star six times as bright as our present North Star.

Fig. 14. Precessional path in heavens described by north end of the earth's axis projected onto the sky. In A.D. 2102, north celestial pole will be near the star that is now at end of handle of Little Dipper. About A.D. 13,500, the north celestial pole will be near bright star Vega.

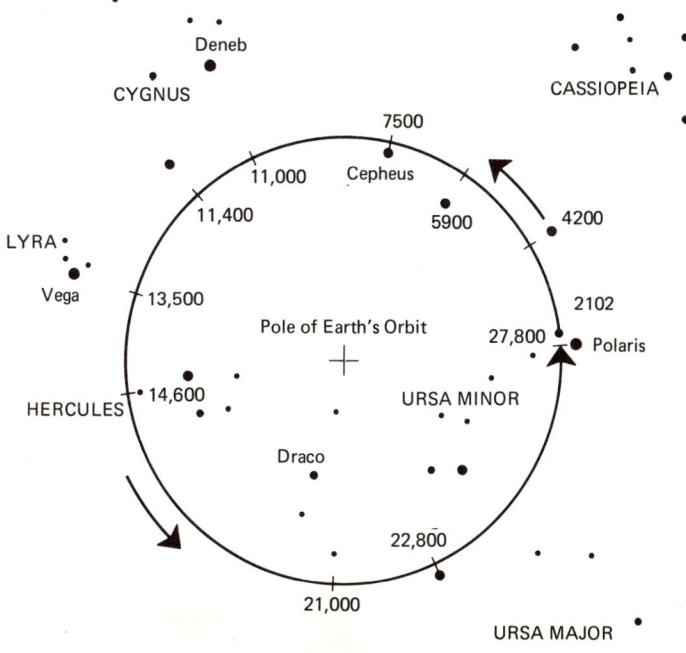

ALEUTIAN ISLANDS

EAST WEST

Coming—The Age of Aquarius

Two thousand years ago, in Hipparchus' time, the vernal equinox was in the constellation of Aries the Ram, which explains the origin of the symbol for the vernal equinox, γ, representing the horns of a ram. Although the vernal equinox has now moved westward into the constellation of Pisces the Fishes, it is convenient to keep the same symbol for the vernal equinox. (Astrologers are a couple of thousand years behind times. They are still reckoning star positions from Aries.) Eventually the vernal equinox will move from Pisces into the next constellation westward, Aquarius the Waterbearer. According to our star map this loudly acclaimed event is some three hundred years in the future, the exact time depending upon where you draw the line between Pisces and Aquarius. If you use the divisions of the zodiac adopted by Claudius Ptolemy about A.D. 139, the Age of Aquarius is 322 years in the future.

Great changes are destined to occur in the thousands of years to come in the aspect of the stars visible at different stations over the earth. Figure 15 shows a view of the midnight sky from the Aleutian Islands on Christmas midnight of A.D. 17,000. Would you guess that the bright stars shining through the aurora are those we now think of as the stars of the southern constellation of Crux, the Southern Cross?

Fig. 15. Precession will produce radical changes in stars seen from different positions on the earth in thousands of years. Thus, about A.D. 17,000, the Southern Cross would be visible near the horizon as seen from the Aleutians.

3 · c = Velocity of Light

"*Sagredo*: But of what kind and how great must we consider this speed of light to be? Is it instantaneous or does it like other motions require time? Can we not decide this by experiment?

"*Simplicio*: Everyday experience shows that the propagation of light is instantaneous . . ."

Galileo Galilei, *Dialogue Concerning the Two New Sciences*, Arcetri, March 6, 1638

One of the most important—probably *the* most important—of the physical constants of nature is the velocity of light. You might suppose therefore that scientists from the earliest times would have exerted themselves to determine its value or at least have done some heavy thinking on the subject. On the contrary, not until the seventeenth century did anyone begin to take an interest in this subject.

Galileo seems to have been the first to make a serious attempt to measure c, the symbol universally adopted for the velocity of light. His method of measurement was basically sound. Two observers with lanterns were stationed less than a mile apart. One observer was to uncover his lantern, and as soon as the other observer saw it, he was to uncover his lantern. With increasing skill, the time between flashes should occur very quickly. (Galileo admitted this probably would require some practice.) I have never been able to find just where and when this famous experiment was performed. In any case, it does not seem to have been what you would describe as a conspicuous success. About all Galileo was able to conclude from his results was that the "velocity of light, if not instantaneous, was extraordinarily rapid." (A statement that today would require a report of a thousand words made out in triplicate.)

And so far as an accurate value for c was concerned, there the matter stood for another century.

Proving the Earth Revolves

"Does the earth move?" you ask the astronomer.

"Yes, the earth moves," the astronomer replies. "The earth revolves around the sun in a period of three hundred sixty-five and a quarter days at an approximate distance of ninety-three million miles."

Most people find it easier to accept the word of authority than to go to the trouble of figuring things out for themselves. But occasionally some nut or crackpot comes along who doesn't see it that way.

"Show me," he demands. "Show me that the earth revolves around the sun."

Now it is fairly easy to demonstrate that the earth has the shape of a sphere and that it rotates on its axis. But it is not at all easy to prove that the earth revolves around the sun. For proof of the revolution of the earth around the sun depends upon minute effects that are difficult to demonstrate and are unfamiliar from everyday experience. Once thoroughly seen and understood, however, they are just as convincing as the proofs of the earth's sphericity and axial rotation.

Suppose for the sake of argument that the earth *does* revolve around the sun once a year in a vast circular orbit 186 million miles wide. Then what observational consequences would we expect from this motion that we could check on?

The first that comes to mind is parallax. What is parallax? Hold a pencil at arm's length in line with a distant tree. Look at the pencil with one eye and then the other. The pencil will appear to shift position back and forth relative to the tree. "Parallax" is the

31

The Stars and Serendipity

technical name for this familiar effect due to the change in your point of view.

If the earth revolves around the sun in accordance with the theory advanced by Nicolaus Copernicus in 1543, you would expect a bright and (presumably) near star to show a slight parallax in the course of a year relative to faint and (presumably) much more distant stars. Great observational astronomers such as Tycho Brahe were well aware of this fact and examined several stars for parallax. Upon finding none they concluded the Copernican theory was wrong and the earth therefore fixed in space.

As time went on, however, evidence accumulated from other sources that supported the Copernican theory, until by the end of the seventeenth century it was generally accepted as an established fact. Astronomers still went on looking for parallax and still were unable to find any evidence for it. The invention of the telescope about 1609 enabled them to measure much smaller angles than was possible with Tycho's old naked-eye instruments. Yet the art of instrument making remained in a pretty bad state. If an astronomer wanted some delicate measuring device to attach to his telescope, about the only person to whom he could apply was the village blacksmith or "armourer." Such an individual was well qualified for pounding out horseshoes but of scant help in constructing a precision micrometer.

Hooke's Chimney Telescope

In 1669 a British scientist (in the seventeenth century they didn't call themselves "scientists," but "philosophers") launched a new attack on the parallax problem. This scientist, or philosopher, was Robert Hooke. There is no question about Hooke's ability as an investigator of great originality and insight. His skill in the construction and manipulation of scientific instruments led to his appointment as keeper of experiments for the Royal Society. He also was possessed of a remarkable talent for making people mad at him. (He nearly drove poor Isaac Newton crazy with his unjustified

c = Velocity of Light

criticism.) Hooke worked on dozens of experiments which would have yielded valuable results if only he had carried them through to completion. But he had an unfortunate habit of starting an experiment and then dropping it for some reason. He started much but finished little.

Hooke set up a telescope thirty-six feet long in his home in London. He chose the oddest place for a telescope—the chimney. Owing to the eastward rotation of the earth the stars always appear to be moving in the opposite direction westward. So to keep a star in view you must turn the telescope westward at the same rate as the stars, otherwise the star passes across the field of view and soon disappears from sight. Now a chimney is not a very flexible type of instrument. How can you possibly keep the stars in view with a telescope lens mounted solidly in a chimney?

But Hooke did not want to observe "the stars." He wanted to observe only one particular star, the star astronomers call Gamma (γ) Draconis in the northern constellation of Draco the Dragon. Why γ Draconis? Because γ Draconis happens to be located at such a place in the sky that it passes almost exactly overhead, or through the zenith, of London. Or putting it the other way around, to an observer in London, γ Draconis passes close to his zenith at some time during every day of the year.

But not the *same* time every day. On June 21, γ Draconis crosses the zenith of London at midnight; on June 22 at four minutes to midnight; on June 23 at eight minutes to midnight, etc., four minutes earlier every night or two hours earlier every month. With a long chimney-type telescope you would have no trouble seeing γ Draconis in full daylight at noon of December 21st.

The advantage of a fixed chimney over the usual style of telescope mounting is its great stability. Instead of turning your telescope to your star you wait for your star to come to you. But, of course, a chimney telescope is useless for observing any object that does not happen to pass through your zenith.

Hooke got his apparatus installed and succeeded in making four observations of γ Draconis. Then disaster struck. He accidentally

The Stars and Serendipity

broke his telescope lens, the vital part of the instrument. Since he never bothered to replace it, the experiment was left unfinished, and the whole effort went for nothing. Had he gone ahead with it he could hardly have failed to make one of the most important discoveries in the history of astronomy.

Bradley Enters the Field

Other astronomers were also preoccupied with parallax. To mention only one, Jean Picard claimed to have measured displacements in the position of Polaris which he attributed to parallax. But his measures showed Polaris moving in such a peculiar way that other astronomers hesitated to accept them as real. Picard also claimed to have measured the parallax, or distance, of a much nearer star—the sun. His value for the distance of the sun was 41 million miles, only about 52 million miles too small. Such was the confused state of the situation when James Bradley, a British astronomer of thirty-one years, entered the field.

Bradley's plan of observation was essentially the same as Hooke's: measures on the zenith distance of γ Draconis made with a chimney telescope. He secured the services of George Graham, one of the foremost instrument makers in the country (Fig. 16).

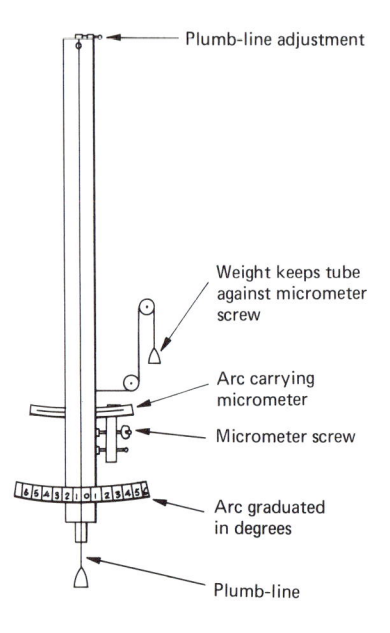

Fig. 16. Bradley's observations of zenith stars with this chimney telescope resulted in his 1728 discovery of the aberration of starlight. The instrument was mounted in the home of his friend Samuel Molyneux, at Kew, England.

c = Velocity of Light

As a result, his equipment was far in advance of any ever used in parallax work before, so sensitive that if γ Draconis had a parallax of no more than a single second of arc, its displacements during the year could not possibly have escaped him.

Bradley's first problem was to find a chimney. One of the drawbacks of converting a chimney into a telescope is that you immediately make it useless for heating purposes. Here he was aided by his astronomer friend, Samuel Molyneux, who granted him permission to use the chimney in his old family mansion at Kew. The apparatus was installed and adjusted and the first serious observation on γ Draconis for parallax made on December 3, 1725.

From the fragmentary information floating around, Bradley could only guess at how much parallax, or shift in position, γ Draconis should show due to the earth's orbital motion. (Assuming it showed any at all!) What he *did* know was how γ Draconis *should* be displaced *if* it did show parallax. Parallax should put the star farthest south of the zenith in December and farthest north in June. Its motion toward the north should be fastest in March and April, and its motion toward the south fastest in September and October. At the extreme limits of its run in December and June it should show little motion either to the north or south.

The *expected* motion of γ Draconis due to parallax is shown in Figure 18. Since Bradley began observing in December he did not anticipate the star would display much evidence of northward motion for probably another month.

In Figure 19 you are supposed to be looking into the eyepiece of Bradley's chimney telescope. We don't know that it looked this way, but for purposes of illustration let's assume that it did. Your field of view is restricted to a width of the sky about equal to one-thousandth the apparent disk of the full moon. The point Z marks the position of the zenith at Kew, determined from the direction of a plumbline. We suppose the eyepiece always adjusted so that the zenith appears at the center of the field.

Imagine you are looking into the eyepiece on a day when some

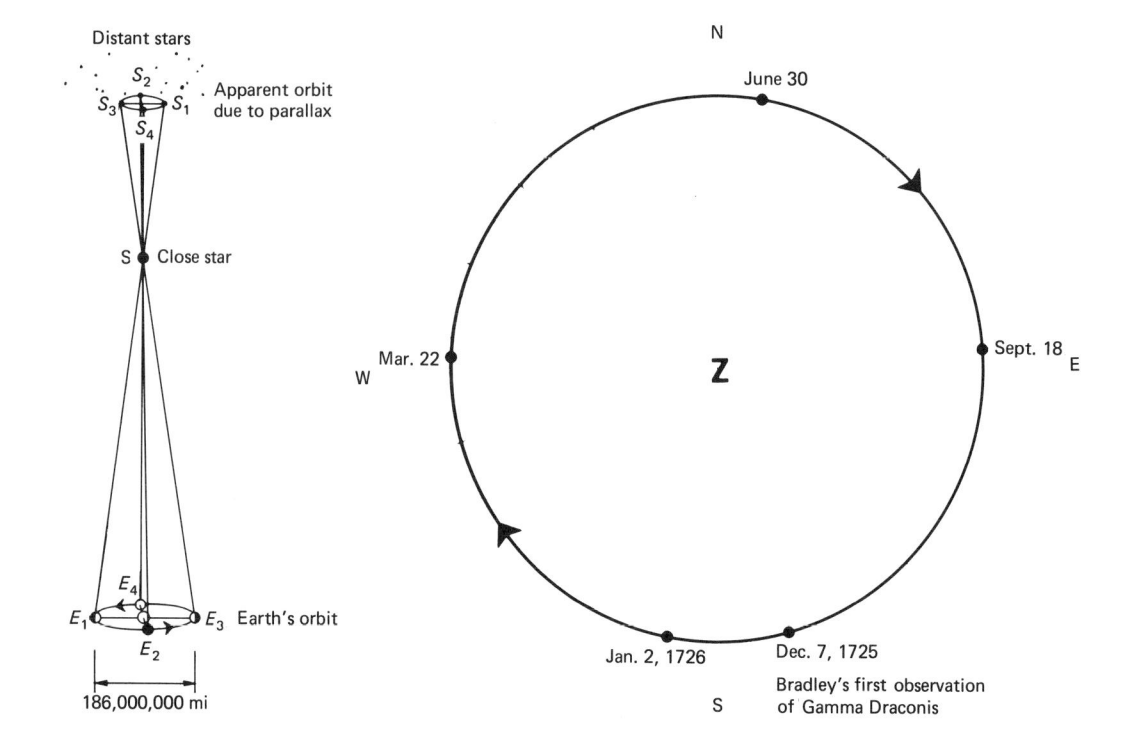

Fig. 17. Above left: The revolution of the earth around the sun causes close star at S apparently to describe small circle seen against background of much more distant stars. About five thousand stars show yearly displacements due to motion of the earth in orbit 186 million miles in diameter. Diagram not drawn to scale.

Fig. 18. Above right: Expected motion of Gamma Draconis due to parallax. Star should be farthest north of zenith in June and farthest south in December.

star passes precisely through the zenith. The star would enter the field at E, move across the field in a straight line from left to right through Z, and pass out of the field at W.

To your eye the stars seem to be moving westward across the sky at a very leisurely rate. But when viewed in a telescope under high magnification their motion seems anything but leisurely. In fact, the star would cross the field from E to W in about twenty seconds. Thus to measure the distance of γ Draconis from the zenith you would want to be prepared well in advance, so you could get to work in a hurry the moment the star shows up. One thing is sure: the star is not going to wait for *you*.

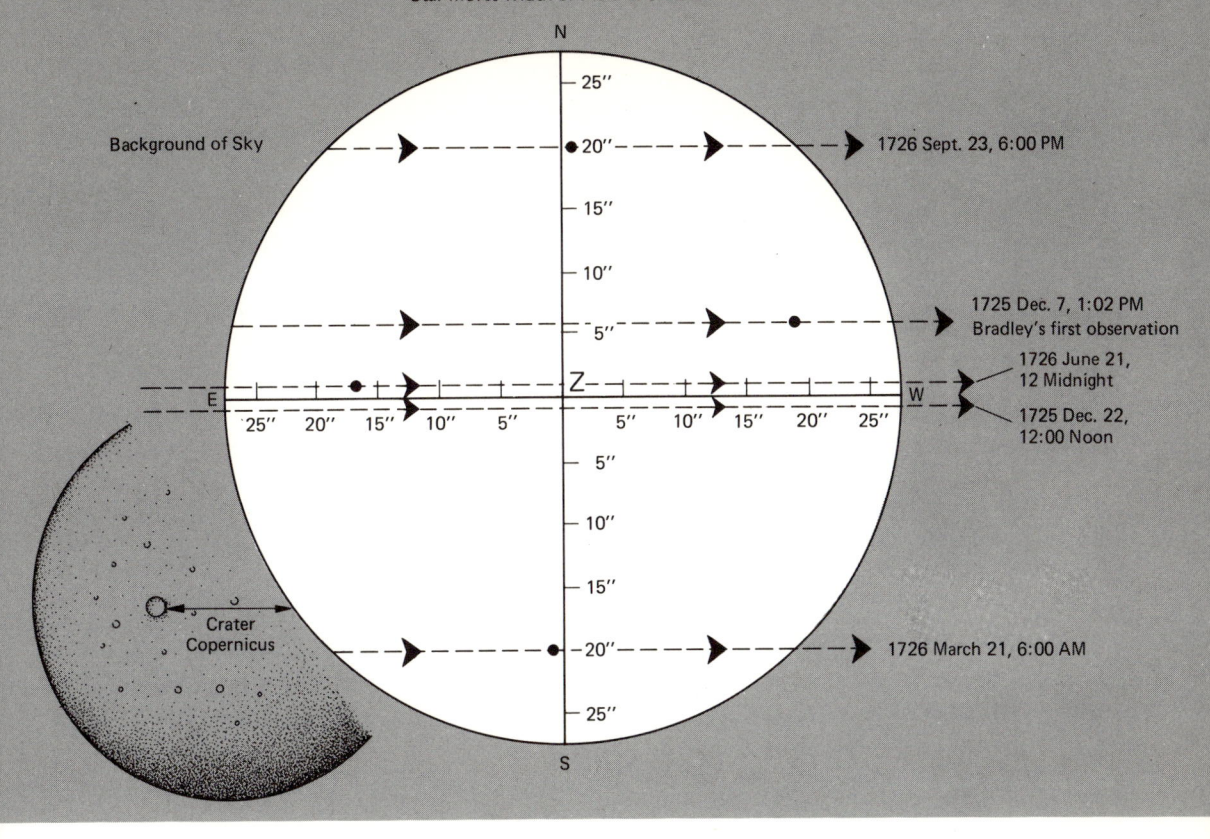

Fig. 19. Field of view looking into eyepiece of Bradley's chimney telescope at Kew. The magnification is assumed to be so high that apparent area of sky would be only about the size of the lunar crater Copernicus (lower left). Z marks the position of zenith at Kew. Gamma Draconis was observed farthest south of zenith in June, and farthest north of zenith in September. Star would enter field on left and move across field to right in about 6.5 seconds.

Bradley first observed γ Draconis on December 3, 5, 11, and 12 of 1725, when parallax should have put it farthest south of his zenith. He seems to have regarded these as dry runs for practice purposes, since no appreciable northward motion was expected till spring. But let's let him tell about it in his own words, as he told it later to Edmund Halley, the Astronomer Royal.

"Noting no material difference in the place of the star, a farther repetition [of the observations] at this season seemed needless, it being a part of the year wherein no sensible alteration of parallax in this star could be expected. It was chiefly, therefore, curiosity that tempted me . . . to prepare for observing the star on

37

The Stars and Serendipity

December 17th, when having adjusted the instrument as usual, I perceived that it passed a little more southerly this day than when it was observed before."

Bradley naturally supposed the displacement south arose from some fault in the instrument or error in his previous observations. What other cause could it be? But upon checking again three days later he found the star even farther south than before. Not only was the star moving when it was not supposed to be moving. It was moving in the wrong direction!

γ Draconis continued its contrary motion until by March 1726, it was twenty seconds of arc (20″) south of its December position. Now an angle of 20″ is not much of an angle. In case you are not familiar with angular measure, you should know that an *angle* is simply the difference in direction between two lines that meet in a common point. This difference in direction may be measured in degrees (°), minutes (′), and seconds (″), of arc. The angular distance measured completely around a circle is 360°. Each degree is divided into 60′, and each minute into 60″. Thus the angular distance around the equator of the earth is 360°. The angular distance from the equator to one of the poles is one-fourth of a whole circle, or 90°. An angle of 90° is often called a "right angle." The angular or apparent diameter of the full moon is very nearly ½°, which is 30′, or 60 × 30 = 1800″. So, as we said, an angle of 20″ is not much of an angle. About the apparent size of a penny one-eighth of a mile distant. But to an astronomer an object out of place by 20″ is a horrifying sight, as horrifying as the sight of a car speeding in the wrong direction up a one-way street would be to us.

It must have been gratifying to Bradley and his associates to find that γ Draconis showed a periodic motion of *some* kind. But what could it be? Whatever it was it could have nothing to do with parallax, for the observed motion always displaced the star in the *same* direction as the earth's motion (Fig. 20). Whereas parallax would displace the star at right angles to the earth's motion (Fig. 21).

Bradley must have asked himself, "Have I discovered something? Do you suppose I've hit on something big? Or have I just got

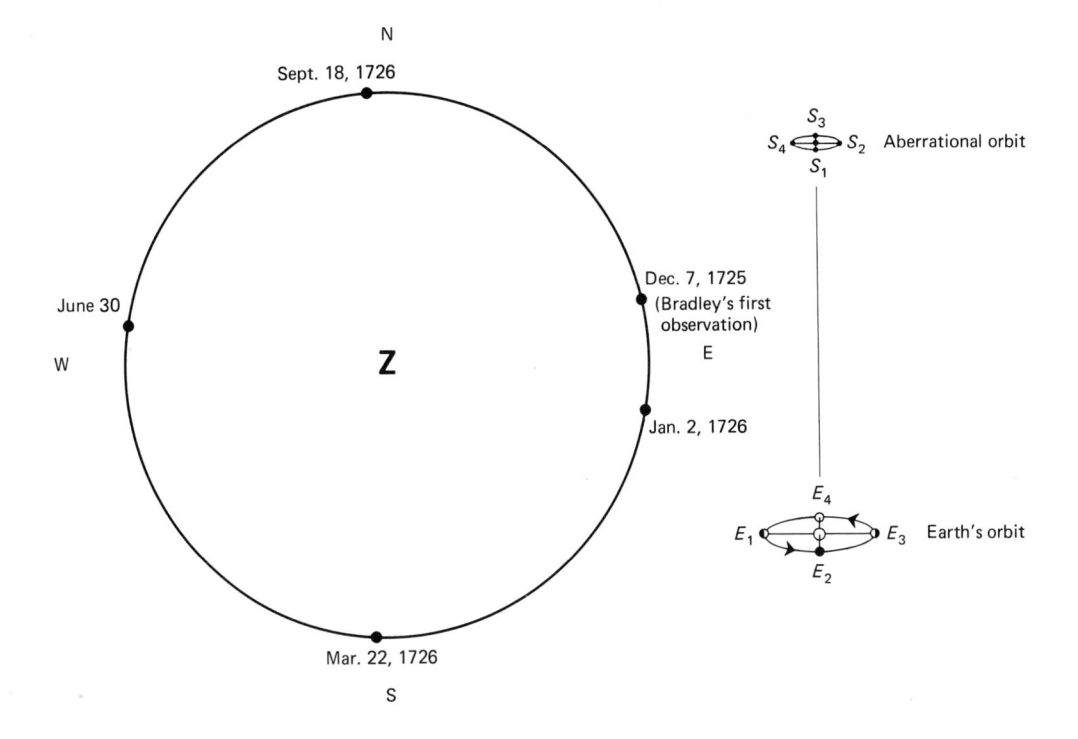

Fig. 20. Above left: Observed *motion of Gamma Draconis due to aberration (unknown at time to Bradley). Star was observed farthest north of zenith in September and farthest south in March.*

Fig. 21. Above right: Aberrational orbit described by star. Star is always displaced in same *direction as the earth's motion. Parallax always displaces star at* right angles *to the earth's motion.*

a bug in my setup somewhere?" In any case, before proclaiming it to the world, he decided it would be a good idea to do some heavy checking first.

So he installed a lens in the chimney of another friend's house with a setup which enabled him to observe stars over a much wider strip of sky than at Kew. (Bradley was lucky in having numerous friends with chimneys to lend.) Besides γ Draconis, he kept close watch on fifty other selected stars. And like γ Draconis, they all showed periodic yearly shifts, although differing in form and amount.

The reality of the effect was now clearly established. But the nature of it remained as much of a mystery as before.

39

The Stars and Serendipity

First Accurate Determination of the Velocity of Light

The story of how Bradley finally got the essential clue to the mystery has been told so often it has come to be accepted as fact. Yet the story never originated with Bradley but in an anecdote in the *History of the Royal Society* published eighty years after Bradley's letter to Halley. According to the story, while crossing the Thames Bradley noticed that the vane on the masthead changed direction every time the boat altered course. When he asked why this happened, the sailors told him that the change in the vane was not due to the shift in the wind's direction, but to the changing course of the boat. Or rather, the apparent direction of the wind as shown by the vane was due to both the direction of the wind *and* the boat.

(It makes a good story but I find it hard to believe that a man of Bradley's intellect would need to have the changes in the wind vane explained to him by sailors on the Thames.)

Application to the stars was clear. Replace the boat by the changing direction of the earth in its orbit. Replace the wind by the steady flow of parallel light rays from a star. By so doing, Bradley was led to the discovery of stellar aberration, the apparent periodic changes which the stars undergo in the course of a year.

We speak of a person who isn't quite right in his head, whose behavior tends to deviate from that of normal people as having an "aberration." Similarly, you might suppose that "stellar aberration" refers to some abnormal behavior of the stars, another piece of technical jargon that has nothing to do with you. But that is where you are wrong. For even though perfectly sane, you have been tangling with aberration in some form all your life. The old raindrop case is still the best example.

It is a stormy day. Although it is raining hard there seems to be no wind, for the drops are falling vertically.

You open your umbrella and start heading north (Fig. 22). You are the earth moving in its orbit. The raindrops are the parallel rays of light from some star meeting the earth's surface. Striding north

Fig. 22. On left, rain is coming from directly overhead when man is standing motionless. On right, when man is moving north, rain seems to be coming from northerly direction.

you discover you were mistaken about the wind: apparently there is a slight wind from the north. At least the drops meet your umbrella *as if* there was a slight wind from the north. Turning west, however, the wind also promptly shifts around toward the west. In fact, whichever direction you go, you always seem to find yourself heading into the wind.

It is convenient to think of the light from a star as consisting of myriads of parallel rays rushing through space at the enormous rate c, the speed of light. The earth moves through these rays at the practically constant rate of 18.5 miles per second. Although the earth's rate of motion changes but little during the year, its direction of motion changes radically. The earth's direction of motion today is opposite to its direction of motion six months ago. And like the rain from the sky, the rays from a star always seem to be changing direction to correspond with the changing direction of the earth's motion in space. The effect is to displace the star slightly in the direction of the earth's motion at the moment. The amount of the displacement depends upon the direction of the star relative to the earth's orbit. When at a maximum, an observer looking at Jupiter would see the center of the planet's disk when actually his telescope is aimed at its edge (Fig. 23).

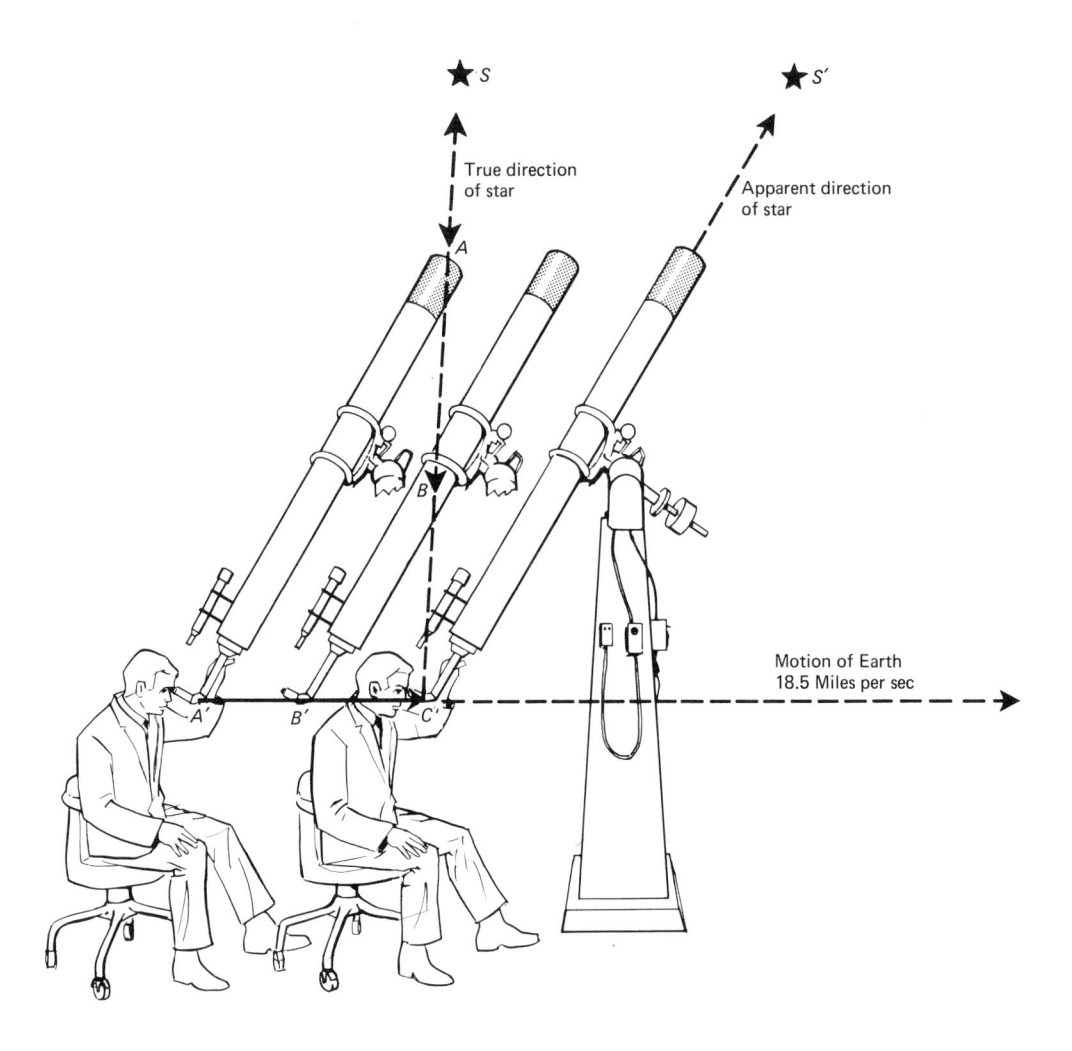

Fig. 23. To see star in center of field of view of telescope, telescope tube must be tilted slightly, owing to aberrational shift. (Tilt is vastly exaggerated in figure.) For example, when aberration is maximum, an observer sees the center of Jupiter when actually looking toward the edge of the planet's disk. From a combination of the earth's orbital motion and observed value of aberration, Bradley derived first accurate velocity of light.

From the known velocity of the earth in its orbit and his measures on stellar aberration—quantities easy to determine—Bradley was able to calculate the velocity of light, c, a quantity otherwise exceedingly difficult to determine. His value thus derived was c = 184,000 miles per second. The velocity of light is so enormous

c = Velocity of Light

that it is no wonder that events on the earth seem to occur instantaneously, as they did to Galileo.

Not until we get off the earth into the vast distances that separate the planets does the "light time," or time required for light to travel these distances, begin to make itself felt. It was from the changing light time in the eclipses of Jupiter's satellites that Olaus Roemer, a young Danish astronomer at the Paris Observatory, was able to get a value for the velocity of light in 1675 of 141,000 miles per second. His value for c was so much in error, however, that it is generally regarded as an "estimate" rather than a "measure."

Today by electronic methods we can determine the velocity of light over distances of a few feet in the laboratory much more accurately than from measures over several miles over the earth's surface or from any celestial measures. From these laboratory methods we get,

$c = 299,793.0$ kilometers per second $= 186,291.4$ miles per second.

We know now that Bradley's original plan for measuring the parallax, or distance, of γ Draconis was hopeless from the start. His chimney telescope was an accurate measuring device for the early eighteenth century but not nearly accurate enough for parallax measures. Not until 1838 did astronomers succeed in developing instruments sufficiently sensitive to detect the minute shifts in star positions due to the back and forth displacement of 186 million miles of the earth in its orbit.

Nevertheless, it was Bradley's carefully planned and executed search for parallax that resulted in his wholly unanticipated discovery of stellar aberration which in turn enabled him to get an accurate measure of the velocity of light. These were major discoveries indeed. For there is much more to aberration than we have described in our simple raindrop illustration. It has important applications in mathematical physics; for example, it is involved in a most interesting way in the special theory of relativity.

Aberration is a classic case of the mysterious operation of serendipity in science.

4 · Uranus and Beyond

". . . the comet appeared perfectly sharp upon its edges and extremely well defined without the least appearance of a beard or tail."
                Account of a comet by Mr. Herschel, F. R. S., 1781
"Train your telescope on the point . . . in the constellation of Aquarius . . . and you will find a new planet . . ."
                Urbain J. J. Leverrier, September 16, 1846

The textbooks will tell you that the planet Uranus was discovered by Sir William Herschel in 1781. There are two errors in this brief statement. Herschel did not know he had discovered a new planet. What he thought he had discovered was a comet. He was not looking for a planet at the time. He was engaged in making a general survey of every object in the sky within the range of his telescope at Bath, England. The idea never seemed to have occurred to him that there might conceivably be another major planet beyond the orbit of the then outermost known planet Saturn. Furthermore, he persisted in calling it a comet, long after it had become obvious to everybody else that it could not possibly be a comet. Also, when he discovered the planet that forty years later became Uranus, he was plain "Mr." Herschel.

Specifically, Herschel sighted this new object on Tuesday, the 13th of March, 1781, between ten and eleven in the evening, in the course of his general survey of the heavens. He was then forty-three, an amateur astronomer, who made his living at Bath as organist at the fashionable Octagon Chapel and as conductor of the Pump Room concerts. There have been plenty of other men who made a name for themselves in astronomy but earned their living in occupations far removed from star study—doctors, lawyers, court reporters, brewers, etc. But Herschel is the only great astronomer who was also a competent professional musician and composer.

Some ten years earlier he had gotten a glimpse of the heavens through a 2-inch telescope that had fired him with an ambition to study the stars that would not be denied. Since obtaining a larger telescope was beyond his resources, he determined to make one himself. This is a fairly easy task today. Hundreds of not particularly skilled people have ground out telescope mirrors of varying quality ranging from 4 to 17 inches in diameter. But two centuries ago grinding a mirror to the proper shape was an awful job. The reason being that the mirrors were not made of glass but of speculum metal, an alloy of copper, tin, brass, silver, and arsenic. Not only is speculum metal hard to grind but there were no power-driven tools to grind them, so that all the labor had to be done by hand. Herschel's devoted sister Caroline said that her brother often ground and polished continuously without taking his hands off the tool for sixteen hours together, while she fed him at meal times. We wonder how Herschel's fingers were in any shape to play the organ after such an ordeal.

After a terrific amount of work and failures that would have disheartened a less dedicated person, Herschel succeeded in producing speculum mirrors of 18.5 and 48 inches in diameter, unheard of dimensions in those days. (His 48-inch mirror does not seem to have been so good. It is said that Herschel never let any astronomer look through it.) Herschel discovered Uranus with a telescope of 6.5 inches aperture and 7 feet focal length, about the same size as the telescopes of the street astronomers of bygone days, who used to show you the moon for a dime.

Herschel distinguished Uranus from the stars simply by its appearance. Although the stars are huge suns vastly bigger than the planets, they are so far away that even in the most powerful telescopes they are reduced to mere bright points. All the planets (except Pluto) show a disk. An experienced observer can generally recognize a planet at a glance. Thus when Herschel spied this object in Gemini he was immediately struck by its "uncommon magni-

> **XXXII.** *Account of a Comet.* By *Mr.* Herschel, *F. R. S.*; communicated by *Dr.* Watson, *Jun. of* Bath, *F. R. S.*
>
> Read April 26, 1781.
>
> ON Tuesday the 13th of March, between ten and eleven in the evening, while I was examining the small stars in the neighbourhood of H Geminorum, I perceived one that appeared visibly larger than the rest: being struck with its uncommon magnitude, I compared it to H Geminorum and the small star in the quartile between Auriga and Gemini, and finding it so much larger than either of them, suspected it to be a comet.

Fig. 24. William Herschel's original announcement of an object he supposed to be comet but actually was the planet Uranus.

tude." That it was within the solar system was confirmed by later observations, which showed it to be moving against the background of the fixed stars. Accordingly, Herschel felt justified in announcing an "Account of a Comet" to the Royal Society on April 26, 1781 (Fig. 24).

On April 6, Herschel had written ". . . the comet appeared perfectly sharp upon its edges and extremely well defined without the least appearance of a beard or tail." The fact that these appendages were lacking is not particularly noteworthy, inasmuch as few comets begin to sprout a beard or tail until within the orbit of Mars. It seems curious that it never occurred to Herschel that his new object might be anything else *but* a comet. We suspect that at that stage in his career Herschel knew more about music than he did about orbits. For it soon appeared that Herschel's comet was pursuing a path quite different from any comet ever known before. Its path, instead of being highly elongated like a cigar, was practically circular. To me it has always seemed that the person who really discovered the *planet* Uranus was the Swedish astronomer Andres Johann Lexell, who first announced that Herschel's object could not be a comet, but must be a "Primary Planet of our Solar System."

Uranus and Beyond

Astronomers immediately began backtracking on Uranus hoping to turn up pre-discovery observations; that is, observations of the planet made unknowingly prior to discovery. Such observations are of vital importance in getting an accurate orbit, especially in the case of a slow-moving body such as Uranus. There proved to be no lack of such observations. So many, in fact, it was hard to understand how Uranus had escaped detection so long.

With pre-discovery observations available extending back nearly a century, and with an abundance of new observations, it would seem that keeping track of Georgium Sidus would soon settle down to a mere matter of routine. (Georgium Sidus was the name Herschel originally proposed for the planet. It is possible that the money and honors bestowed upon him by his gracious majesty, George III, had something to do with it. Needless to say, his proposal was received with scant enthusiasm outside of England.) On the contrary, Georgium Sidus gave astronomers trouble right from the start. They could derive an orbit that fit the old observations. They could derive an orbit that fit the new observations. What they could not do was to derive an orbit that fit *both* the old *and* the new observations.

You must not suppose that Uranus went wandering around the sky in imminent danger of getting lost. The difference between observation and theory at most never amounted to more than one-fifteenth the diameter of the full moon, an angular separation hardly perceptible to the unaided eye. But to the astronomers it was intolerable.

The astronomers of the early nineteenth century who labored so industriously trying to keep Georgium Sidus under control must have realized that it was being disturbed by the attraction of a still more distant planet. (The name Uranus was first suggested in 1823, but Georgium Sidus was retained by the British Nautical Almanac Office until 1850.) But the old established astronomers stubbornly resisted such an idea. For some reason they were curiously reluctant to admit the existence of any planet in the solar system not already known to them.

The first person bold enough to express such a radical notion was

The Stars and Serendipity

the Reverend T. J. Hussey, an English amateur. (Until the twentieth century practically all astronomers were "amateurs." Who would pay anybody for such a useless occupation as looking at the stars?) In a letter to Sir George Airy, a year before his appointment as Astronomer Royal, Hussey volunteered to conduct a search for such an unknown body with his own telescope, provided Airy would furnish him with an approximate estimate of where to look. Airy replied that he still was not convinced that the deviations of Uranus from prediction did not arise from errors in the theory. Besides, even if another planet were pulling Uranus out of position, he doubted if it was possible to estimate its position. And so the Reverend Mr. Hussey's offer went for nothing.

The stage was set for some vigorous young astronomer to tackle Uranus, unhampered by preconceived notions regarding its behavior. One such person was John Couch Adams, a brilliant mathematics student at St. John's College, Cambridge, England, who in 1841 at the age of twenty-two was already planning to investigate Uranus "as soon as possible after taking my degree." Another was Urbain J. J. Leverrier, across the Channel in Paris, eight years older than Adams and entirely unknown to him. Leverrier started his career with a treatise on the chemistry of smoking tobacco and only got into astronomy later.

Adams was the first to finish. In September 1845, he sent a letter to Professor James Challis, director of the Cambridge Observatory, giving his orbit for the hypothetical planet with information on where to look for it. On October 31, he sent the same information to Airy, the newly appointed Astronomer Royal. Airy had been investigating the motion of Uranus and hence was thoroughly familiar with the problem. Adams' predicted position fell in a comparatively vacant region of the sky in Aquarius far outside the Milky Way, where search would not be hampered by a multitude of star images. A careful examination of the star images around Adams' predicted position could hardly have failed to reveal an object with a disk as large and bright as Neptune.

Uranus and Beyond

Now you would be surprised how many people write letters to astronomers proclaiming the stupendous discoveries they have made, and demanding that something be done about them immediately. Such letters are generally acknowledged with a polite note of thanks and then consigned to oblivion in the wastebasket. But Adams' letter was obviously not in the crank or crackpot class. It must have been evident at a glance that it was written by a man who knew what he was talking about. You would naturally suppose that Airy's interest would have been aroused at once. Instead he delayed. He wrote Adams requesting additional information on a technical point. Adams for some reason did not answer. Some four months passed. By that time the sun had moved so near the planet as to make the search for it impossible—all of which furnished an excellent excuse for dropping the matter.

Meanwhile Leverrier, wholly unknown to Adams, had been busy across the Channel. He sent a communication on his results to the French Academy on November 10, 1845, and another on June 1, 1846. His published results agreed so closely with those of Adams that Airy was finally aroused to action. But first he had to make an estimate of how many hours would need to be devoted to the search. This important consideration having been settled he finally told Challis to start work at the telescope.

Challis' quickest method of search would have been to try to distinguish the planet from the stars simply by its appearance, the same way that you pick out a friend from among a crowd of strangers. The stars are so distant that they always look like points of light, and always remain points of light, no matter how much you magnify them or how big your telescope. But all the planets are near enough, with the single exception of Pluto, to show a disk. Jupiter to the eye appears as a bright yellow star. But even in a 2-inch spyglass under a magnification of 30, Jupiter presents a distinct disk. You never would mistake it for a star. To see the smaller and more distant planet Neptune as a disk requires considerably more optical power, a telescope with a lens of about 20 inches and a magnification of 300.

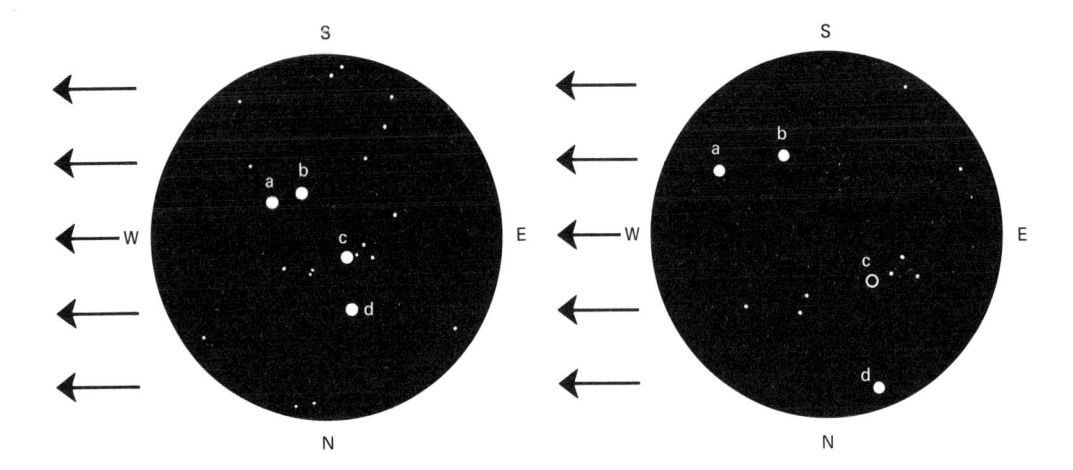

Fig. 25a. Above left: Star field of telescope under low power. One of stars a, b, c, d is suspected of being a planet. If telescope driving clock were stopped all stars would immediately appear to be moving west as shown by the arrows, due to the rotation of the earth. Driving clock makes star apparently stand still. Directions are as seen in inverting telescope.

Fig. 25b. Above right: Same star field as seen under high power. Stars are spread farther apart but relative positions are unchanged, except for c which now shows small disk, thus identifying it as the planet.

Figure 25a shows the stars as seen in the field of view of a large telescope under low magnification. (You can readily change the magnification of a telescope by changing the eyepiece in the end.) One of the four bright objects—a, b, c, and d—is suspected of being a planet. But so far as appearances go they all look pretty much like stars. So we take out the low-power eyepiece and insert one of much higher power. The high power reveals the planetary character of c immediately (Figure 25b). The other three stars appear essentially as before. The only effect of using higher magnification is to spread them farther apart.

Challis tried to detect the planet not by its appearance but by its motion. This method would not have been particularly difficult had he possessed a good star map of the region in Aquarius where Adams' calculations placed the planet. But lacking a star map he had to make his own. He would map the stars in a certain

Uranus and Beyond

small region, then go back a few days later and see if any had moved. If the planet was there, the method was sure to reveal it in time—provided you had enough time.

Leverrier, on the other hand, got quick action on the observational end. On September 16, 1846, he wrote to John G. Galle at the Berlin Observatory, "Train your telescope on the point of the ecliptic in the constellation of Aquarius (longitude 326°) and within a degree of that point you will find a new planet, looking like a 9th magnitude star and showing a small disk." (Nobody could ever say that Leverrier lacked confidence in himself. He had an arrogant, imperious disposition that made it impossible to get along with him.) Doubtless he sought help in Germany rather than his own country, because he had heard that the Berlin Observatory was preparing star charts of this region. Sure enough! H. L. d'Arrest found a chart covering just that part of Aquarius desired.

That evening Galle and d'Arrest began comparing stars on the chart with those of about the right brightness they observed in Aquarius. A brief search revealed a star somewhat brighter than Leverrier predicted where none appeared on the chart. Was it moving? They kept the suspicious object under observation until after midnight but without being able to detect any certain displacement. But the following night there could be no further doubt: the object had definitely moved. And just as Leverrier had predicted, it was within less than a degree of the point designated. Official date of discovery was September 23, 1846. As it happened, Adams' predicted position was a trifle closer to Neptune's observation than Leverrier's (Fig. 26).

The discovery of Neptune was a spectacular feat which caught the imagination of scientists and public alike. Yet it is doubtful if Adams or Leverrier regarded it as of much importance. Adams'

Fig. 26. Overleaf: Leverrier's and Adams' predicted orbits for Neptune based upon irregularities in the motion of Uranus. Notice that although both orbits depart widely from Neptune's true orbit, yet their predicted direction for Neptune is very nearly the same as Neptune's actual direction.

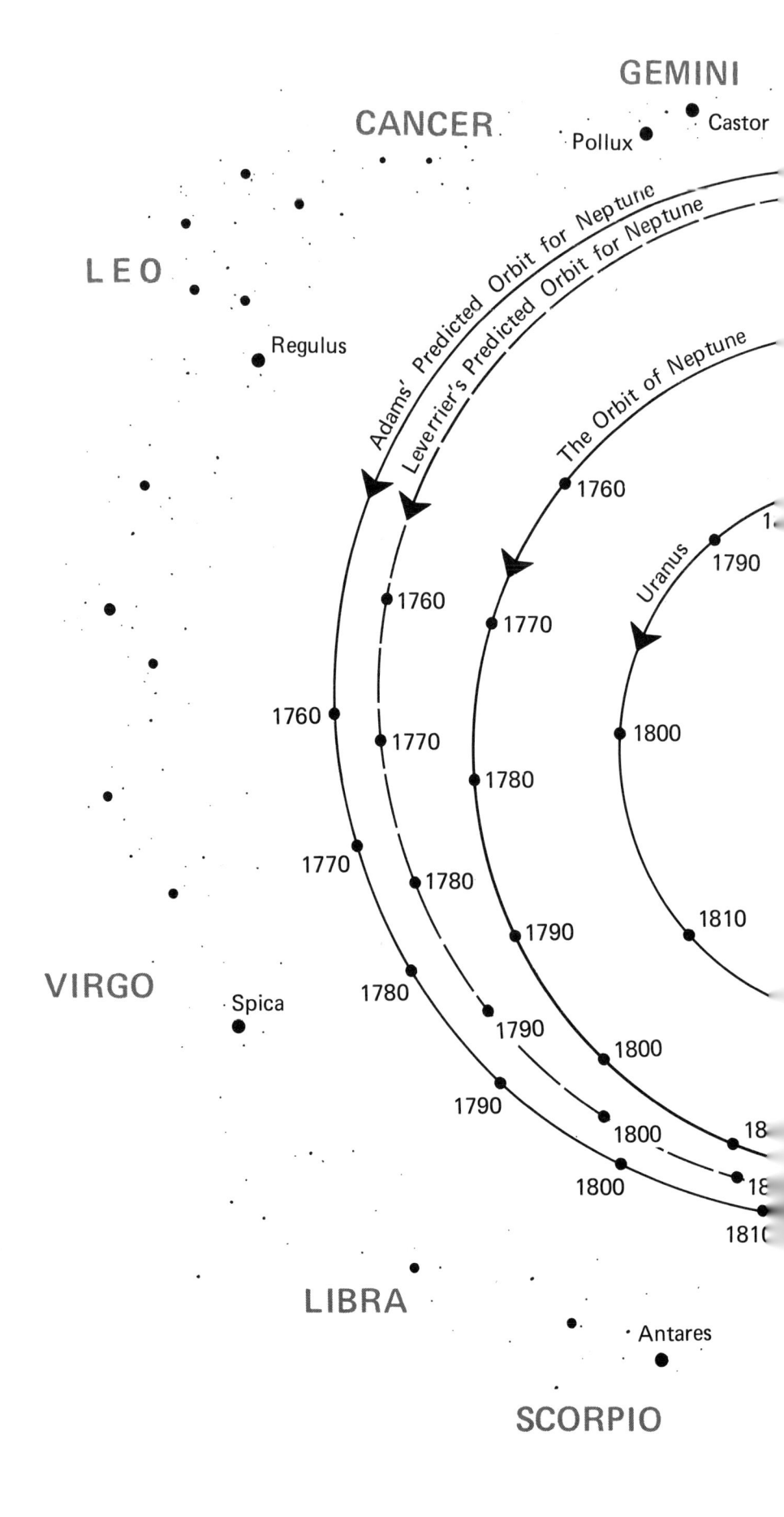

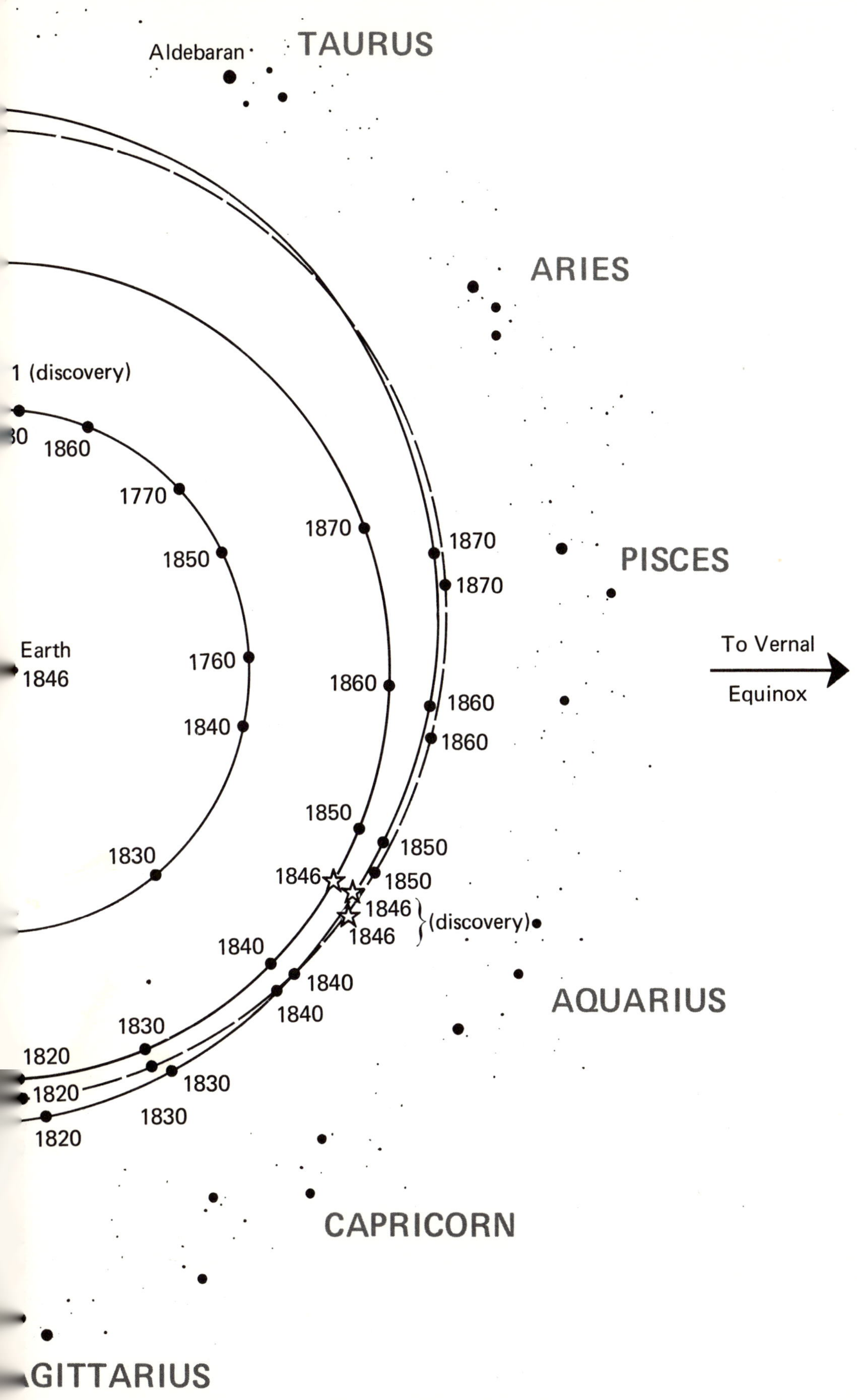

TAURUS
Aldebaran
ARIES
PISCES
AQUARIUS
CAPRICORN
GITTARIUS
To Vernal
Equinox
1 (discovery)
Earth
1846
1860
1770
1850
1760
1840
1830
1870
1860
1850
1846
1840
1830
1820
1870
1870
1860
1860
1850
1850
1846
1846
(discovery)
1840
1840
1830
1830
1820
1820

The Stars and Serendipity

later investigation of the motions of the moon which resulted in the discovery of the acceleration in the earth's rotation seems of much more significance now. And to Leverrier, the discovery of Neptune was merely an incident in his development of a general theory of planetary motion.

As we remarked, when Herschel began his systematic survey of the heavens he hadn't the faintest expectation of finding a new major planet. Still less could he have anticipated that discovery of his "comet" would ultimately point the way to another major planet. Even that was not the end. For after the discovery of Neptune, Uranus continued to deviate from prediction. And it was these slight deviations that resulted in the discovery of Pluto in 1930.

The discovery of Uranus was pure serendipity. The discovery of Neptune and Pluto was the result of prediction from theory and a lot of hard work. Some discoveries in science don't come so easily.

5 · Sunspots

"Solar spots observe no regularity in their shape, magnitude, number, or their time of appearance or continuance."

Long's *Astronomy*, 1764

Edgar Allan Poe in one of his lesser-known tales, *The Man of the Crowd*, speaks of a German book of which it was said, "It does not permit itself to be read." Meaning that it contains certain secrets of so dreadful a nature that they do not permit themselves to be revealed. Similarly, in science you might say that there have been certain discoveries which did not permit themselves to be predicted. The rings of Saturn are an example. The discovery of spontaneous radioactivity is another.

The existence of large, dark spots on the sun was equally unpredictable. It is safe to say that no scientist today would believe there are dark spots on the sun unless he could see them. Who should have the honor of this discovery is uncertain. It is said that Johannes Fabricius first observed them in 1610—not with a telescope but with the camera obscura (a darkened chamber in which an image of a bright object is formed on one wall by admitting light through a small hole). In any case, it was not much of a discovery. Anyone with a small telescope can easily see the spots that come and go on the solar disk. And there are records of the naked-eye sighting of huge spot groups on the sun centuries before the invention of the telescope (Fig. 27).

"Solar spots observe no regularity in their shape, magnitude, number, or in the time of their appearance or continuance."

Judging from this remark culled from Long's *Astronomy* of 1764, it is improbable that any astronomer of that period regarded sun-

Fig. 27. Occasionally large spot groups may be visible to the unaided eye when sun is seen at sunrise or sunset through thin haze.

spots as likely objects for a crash observation program. The moon, sun, and planets got a lot of attention owing to the considerable regularity of their motions. But sunspots . . . What could you do with them?

In the two centuries following the invention of the telescope there were men who observed the sun and kept regular records of the spots. But always after a few months or years they lost interest in them and turned to what seemed more rewarding fields. There is an old saying, "The more things change, the more they remain the same." And so it seemed with the solar spots.

Sunspots

If on Sunday, October 30, 1825, you had been in Dessau, Germany, you would have noticed that not everybody in the village was in church or asleep. For there was evidence of some kind of activity on the rooftop of Samuel Heinrich Schwabe, the local apothecary. Schwabe had recently acquired two telescopes by Fraunhofer, the foremost optician of his day. They contained lenses of about 3 inches aperture, scarcely more than mere spyglasses. But in 1825 they would have rated as fairly respectable telescopes, and must have represented a sizable investment. It is hard to believe today that in 1843 the largest lens type of telescope in the United States, and the second largest in the world, was the 11-inch glass of the Cincinnati Observatory.

At the suggestion of a friend, Schwabe had started a program for scanning the sun's disk in the hope of finding a small planet revolving within the orbit of Mercury. His best chance of finding such an object would be in catching it when in transit across the sun's disk, when it would appear as a tiny black dot silhouetted against the bright background of the sun, easily distinguished from the solar spots by its motion.

Others had looked for intra-Mercurial planets without success, and Schwabe had no better luck than those before him. But he soon forgot all about intra-Mercurial planets in his fascination with the ever-changing spots. We said that other observers soon tired of them. Not so Schwabe. For the next thirty years the sun never shone over Dessau without being confronted by Schwabe and his telescope.

Apparently Schwabe kept a record of the spots visible each day by projecting the image of the sun on a sheet of paper and making a tracing of the different spot groups. Schwabe did not care to publish anything until 1838 when he had a complete series of observations extending over thirteen years. Although his counts showed some indication of periodicity, he was content simply to announce his results without attempting to use them as a basis for a new theory of the universe. (In contrast to the methods of some astronomers today.) Finally, in a note to the scientific journal *Astronomische Nachrichten* in 1843, he remarked that his observations since

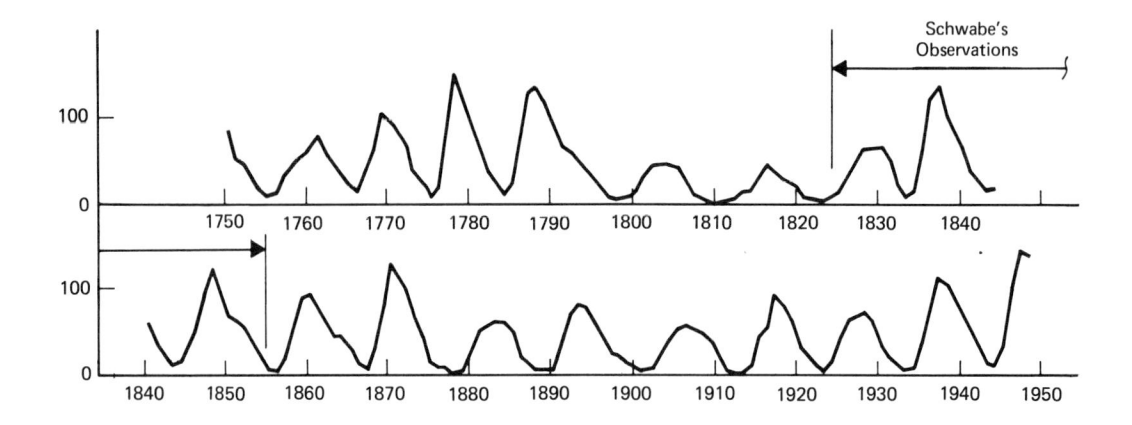

Fig. 28. Smoothed curve showing variations in rise and fall of number of spot counts. Notice that Schwabe's observations covered two maxima of activity. Sunspot curve previous to Schwabe is uncertain. Vertical scale is arbitrary index of sunspot activity determined at Zurich.

1826 indicated a variation in the number of spot groups in a cycle of about ten years (Fig. 28). He buried this vital bit of information in the end of his paper where people were least likely to read it. And apparently nobody *did* read it. It attracted no attention whatever.

The sunspot cycle might have had to be discovered all over again had not Schwabe himself been discovered. (This sort of thing occasionally happens in science. The hazy patch of light opposite the sun called the "Gegenschein" had to be discovered three times before its existence was generally recognized.) Somehow the work of this obscure apothecary of Dessau came to the attention of Alexander von Humboldt, the world-renowned explorer and writer. Humboldt's interest ranged over the whole field of science, from astronomy and biology to volcanoes and zoology. And so when the fourth volume of Humboldt's *Cosmos* appeared it contained a full account of the observations of the sun made by Schwabe. Whereas circulation of the *Astronomische Nachrichten* was restricted to the scientific elite, Humboldt spoke to a wide audience. The result was that announcement of the ten-year sunspot cycle* struck the world

---

*Now known to be eleven years on the average. The sunspot cycle has no definite length.

with all the impact of a fresh discovery. In fact, except for the discovery of Neptune in 1846, it was about the *only* genuine discovery of any importance that had been made for years in the supposedly "exhausted" subject of astronomy.

Belated recognition came to Schwabe in 1857 when he was awarded the gold medal of the Royal Astronomical Society. The depressing state into which astronomy had fallen can be judged by the speaker at the presentation ceremony who remarked, "I can conceive few more unpromising subjects, from which to extract a definite result, than were the solar spots when Schwabe first attacked them."

Schwabe had evidently never heard of serendipity. Instead he drew an example from the Bible, comparing his wholly unanticipated discovery of the sunspot cycle to the story in the Old Testament of Saul, who set forth on the lowly errand of looking for his father's donkeys, and had his mission end by being crowned a king.

6 · The Strange Companion of the Dog Star

"I do not see how a star which has once got into this compressed state is ever going to get out of it . . . It would seem that the star will be in an awkward predicament when its supply of sub-atomic energy ultimately fails. Imagine a body continually losing heat but with insufficient energy to grow cold!"

A. S. Eddington, *The Internal Constitution of Stars*, 1926

Anyone who is aware of the stars at all must know Sirius the Dog Star, the brightest star in the sky. (Except, of course, for the sun.) On winter nights we who live in the United States see Sirius sparkling in the south like a diamond. That outworn phrase "sparkling like a diamond" is the only one that fits. To me Sirius looks exactly like a big blue-white sparkling diamond.

Actually Sirius is not intrinsically a particularly luminous star, about twenty-three times as luminous as the sun. Rigel, another bright-white star nearby, is forty thousand times as luminous as the sun. Sirius appears bright only because it is so close, the sixth nearest the sun, a mere 8.7 light years away. (1 light year = 5,880,000,000,000 miles). Sirius is not a single star but a stellar system, consisting of the bright star we see and a close, faint star we can only detect under special observing conditions. The bright star we see is not an especially interesting object. Much more interesting is its faint companion that we *can't* ordinarily see. There are times when you can see this companion with a telescope as small as a 6-inch, then for years it will be out of reach or observable only with great difficulty even in the most powerful telescopes.

The fact that Sirius is a double star, or "binary," is not in itself unusual. Binary and multiple systems may outnumber isolated stars such as the sun. That Sirius has a faint companion was con-

The Strange Companion of the Dog Star

fidently predicted nineteen years before it was seen. Such a prediction was not so remarkable, either. What no one by any stretch of the imagination could possibly have predicted was the peculiar nature of that companion.

Consider the usual objects on top of a desk. There may be a vase filled with flowers, some pencils, a magnifying glass, etc. On my desk I have a meteorite the size of an egg that weighs about a pound. It makes a good paperweight.

What happens if you leave these objects alone?

Why, nothing will happen, naturally. They will stay right there where they are now. They won't move unless you apply some force to make them move. Like bumping into the desk, for instance. A good hard bump might tip over the vase, knock the pencils onto the floor, and move the magnifying glass by several inches. The meteorite, being so heavy, would scarcely be displaced at all.

Let's make the question a little harder. Suppose the body is already in motion. What happens if you leave it alone?

What's so hard about that? It would come to rest pretty soon, too. A player kicks a football. The force applied by his toe sends the ball soaring. It rises higher and higher, stops, begins to fall, hits the ground, bounces around, and comes to rest. Friction and contact with other bodies soon bring everything to rest. To keep a body moving you've got to keep applying force to it. Give the ball another kick or scoop it up and run with it.

Suppose the body is a rocket far above the earth, in a vacuum, propelled by its exhaust. The exhaust goes off. What would happen to the rocket?

That would depend upon conditions. If a rocket in a vacuum is moving fast enough, it will continue revolving around the earth indefinitely.

"But if the exhaust is no longer operating, what is there to keep the rocket moving?" people demand.

Well, there is nothing to keep it moving. It doesn't need anything. A better question would be to ask, "What is there to stop it from

The Stars and Serendipity

moving?" In a vacuum there is nothing to stop it from moving. No friction or contact with other bodies. So it just keeps on moving.

This is so contrary to everyday experience that often people are still not convinced. They can't understand how a body can continue moving unless you give it a shove or kick now and then. Yet they are used to seeing the moon up there in the sky. Where is the force that keeps the moon moving?

Some three hundred years ago Isaac Newton gave a general answer to questions such as these. He first arrived at these answers when he was twenty-three. They are contained in the three laws of motion as set forth in his *Principia*. Only the first law concerns us here. It may be stated as follows:

"Every body continues in a state of rest or uniform motion in a straight line unless compelled to change that state by the action of some force upon it."

This simple law is worth memorizing and thinking about. What Newton saw in a flash of genius is that being in motion is just as natural a state for a body as being at rest. If a body is in motion, and no force is acting to change its motion, it will continue in motion.

(One question Newton never answered in the *Principia* was why he waited twenty years to publish his results.)

Now back to Sirius...

Sirius appears to be a star moving alone through space. If alone, it should move in a straight line at uniform speed. But observations show Sirius is moving in a wavy line at a non-uniform speed. Hence, some force must be acting upon Sirius to change its motion.

But what can this force be? Where does it originate?

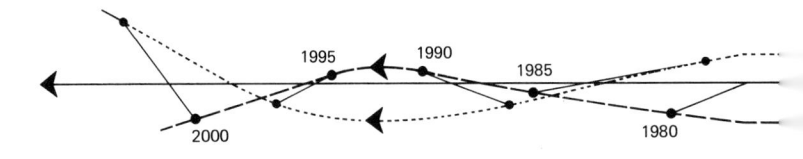

The Strange Companion of the Dog Star

The only possible answer is that Sirius is *not* alone in space. Sirius *must* have a companion—but a companion too faint to be visible. It is the gravitational attraction of this companion that is pulling Sirius out of its straight-line path.

It was F. W. Bessel, a German astronomer, who first suspected that Sirius was deviating from a straight-line path. His suspicions were aroused as early as 1834 from comparison of the various measures made on Sirius since 1755. With commendable scientific caution, he deemed it advisable to delay announcement for another ten years until more observations were available. By 1844 he felt justified in announcing to the Royal Society that "stars whose motions since 1755 have shown remarkable changes, must (if the change cannot be proved to be independent of gravitation) be parts of stellar systems" (Fig. 29).

Fig. 29. Lower curved lines show absolute orbits of Sirius B (dotted line) and Sirius A (dashed line). Straight line below shows motion of center of gravity of Sirius system. Oval curve, upper right, shows orbit of companion relative to bright star A.

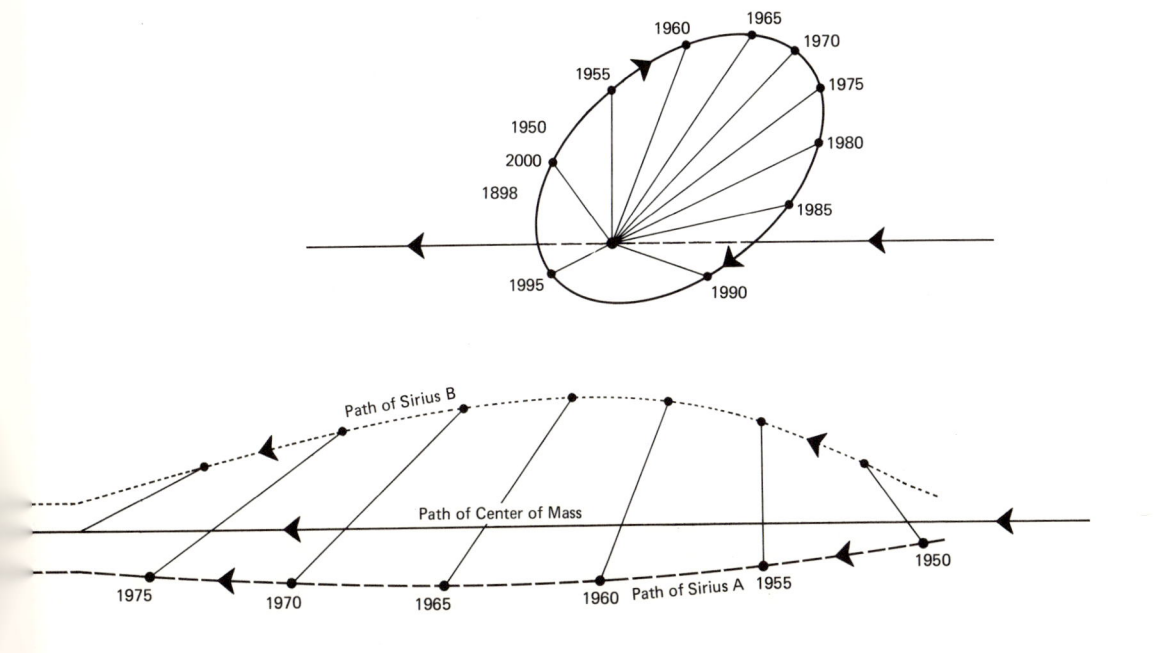

The Stars and Serendipity

Bessel was one of the foremost astronomers of the nineteenth century. Besides being the first to measure the parallax of a star, he also made important contributions to mathematics. Yet in his youth he hadn't planned to be an astronomer. He started life as a merchant.

In the twenty years following Bessel's announcement the wavy motion of Sirius furnished several leading astronomers with a splendid opportunity for displaying their mathematical talents. The problem was to invent an orbit for an invisible star, Sirius B, that would account for the observed irregularities of the visible star, Sirius A. The problem was not completely determinate. The observed motion of Sirius enabled them to predict the *direction* of B from A, but not the *distance* of B from A. It was this uncertainty in the distance that may explain why astronomers became discouraged and ceased looking for B.

Discovery

Until the middle of the last century the best lens telescopes, or refractors, came exclusively from Europe. (They are called refractors because the glass lens in the outer end bends, or "refracts," the light to a focus.) The laws governing the grinding of lenses to the proper shape were well-known to opticians. Yet that does not mean that every optician could apply them to turn out first-rate telescope lenses. Lens-making is not a straightforward operation like laying a cement sidewalk. It is an art rather than a craft. And for a long time there was no one in this frontier country who possessed this art.

About 1846, a portrait painter of Cambridgeport, Massachusetts, Alvan Clark, began making small telescopes as a sideline. Soon he was producing lenses of the highest quality, but as so often happens, his own countrymen were the last to realize the fact. Not until 1853 did he receive recognition, when the Reverend W. R. Dawes, a noted British amateur, purchased several of his lenses which he endorsed in the most enthusiastic terms. Perhaps it was from Clark's lenses

Fig. 30. Photograph of Sirius and companion with the Sproul 24-inch refractor. The six-sided diaphragm on the camera scattered light to make the image of the brighter star appear to have six points. March 26, 1961.

that Dawes secured his surname of the "Eagle-eyed." In 1860 the firm of Alvan Clark & Sons received an order for a lens 18 inches in diameter, exceeding by 3 inches the largest telescope lens then in existence.

On the evening of January 31, 1862, Clark and his son were giving their new 18-inch a preliminary test by examining the quality of the image formed of different objects. Turning the telescope on Sirius, the experienced eye of the younger Clark at once noticed something unusual. Due east of the bright star, and almost drowned in its brilliance, was a faint star he had never seen before (Fig. 30). He called his father who confirmed its presence. Doubtless they took precautions to make sure they were not looking at a "ghost," a false image that looks so real it often fools experienced observers who should know better. When discovery of the faint companion was announced, other astronomers, as so often happens, had no trouble seeing it too. Moreover, its direction from Sirius was exactly in accordance with theory. (The Clarks had no prior knowledge of the theoretical aspect of the problem.)

When conditions are right, you can discern the companion of Sirius with a telescope of 6 inches aperture. Many astronomers after Bessel's announcement of 1844 possessed telescopes exceeding 6 inches in aperture. Why then was the discovery of the companion of Sirius so long delayed?

The Stars and Serendipity

The Mystery Explained

You can see the answer back in Figure 29. The diagram shows the path of B and A relative to faint stars in their vicinity. A is ten thousand times brighter than B. If you have ever been on the stage blinded by the glaring lights shining in your face you will understand why B is so hard, or impossible, to see in certain parts of its orbit. If near A, it is going to be obliterated. B revolves around A in a period of fifty years. During about twenty of these years, B can only be detected with great difficulty, if at all.

It is worth emphasizing that we would have no trouble seeing B were it not for the presence of A. It is the glare of light from A, scattered and diffused by our atmosphere and the lenses and mirrors in our telescopes, that blots out the companion. Remove Sirius A and you could see Sirius B with a 2-inch spyglass.

Figure 29 shows positions of B relative to A as if the Sirius system were fixed in the sky. But we know that the Sirius system—the two components A and B—have a common motion in space which in 1,400 years will displace the Dog Star by the apparent width of the full moon. The motion of the center of mass of the system is shown by the straight line in Figure 29. The wavy lines twining about the straight line indicate the motion of A and B relative to their common center of mass. Notice that B swings about twice as far from the straight line as A. This is because A is about twice as massive as B. It is like a 100-pound boy and a 200-pound man on a teeter-totter. The boy does most of the moving.

Many astronomers have worked on the Sirius system and, as you would expect, have derived slightly different masses for the two components. Here we shall adopt the values:

    Mass of Sirius A . . . . 2.17 × mass of sun
    Mass of Sirius B . . . . 1.02 × mass of sun

You can see from the diagram the difficulty in detecting Sirius B during certain periods. Now, of course, knowing the motion of the faint companion we can predict approximately when it should emerge into visibility.

If this were all there were to the story of the Sirius system it would be merely another example of the accuracy of prediction from gravitational theory. But we have not come to the interesting part yet. The part that, a century ago, no one in his right mind could possibly have predicted from theory.

Astronomers Discover the Stars

Soon after Sirius B was first sighted, astronomers made another big discovery—*the stars*. For centuries they had been concerned scarcely at all with the stars as *physical bodies* in the same way that they were concerned with the physical nature of the moon, Mars, Jupiter, and Saturn. Rather, their chief concern was with the positions and motions of the stars as points of light on the celestial sphere. But late in the nineteenth century a few hardy pioneers began asking, with increasing zeal, such questions as "Why are the stars of different colors? What chemical elements are they made of? What makes them shine? How much longer will they continue to shine?" And many others of an equally embarrassing nature.

Previously there had been little point in raising such questions since the prospect of getting answers seemed hopeless. But about 1870 there came a glimmer of light. A few astronomers began attaching a new-fangled gadget called the spectroscope to the end of their telescopes. The spectroscope was destined to revolutionize astronomy but at the time it was regarded with distrust by most members of the profession. To understand what follows we shall need at least a nodding acquaintance with this instrument.

What Is a Spectroscope? What Does It Do?

The astronomer would be unable to carry on his work without the spectroscope. Take all the spectroscopes out of the observatories and the astronomers would practically have to close up shop. Yet the average person, and even many scientists, have never encountered a spectroscope—which makes it rather difficult to talk

The Stars and Serendipity

about this really simple instrument because it is totally unfamiliar to most people from firsthand experience. You don't need a spectroscope to operate your dishwasher or start your car.

The most familiar example of a spectrum is the rainbow, the colored band arching overhead produced by sunlight shining through a sky filled with raindrops. Everyone has also noticed the flashes of color from a crystal chandelier. In a chandelier the glass is deliberately cut in different shapes to jumble up the colors as much as possible.

In a spectroscope the glass is cut in the shape of a three-sided prism with the angles between the sides nearly equal. A beam of white light falling upon the faces of one of the prisms at the proper angle will emerge as a continuous strip of rainbow colors. Such a spectrum is not very good due to overlapping of the colors. We can get a much "purer" spectrum by admitting the white light through a narrow slit, and by inserting lenses of suitable size and curvature in front of the prism and behind it.

If we catch the light coming from the prism on a white card, we will see a strip of colors merging insensibly into one another, from red at one end through orange, yellow, green, blue, indigo, and violet at the other (Fig. 31). Beyond the red and violet we are unable to see any color or light whatever.

Fig. 31. Crude formation of solar spectrum when a beam from the sun is passed through a round hole into a prism and the colors are caught on a white card.

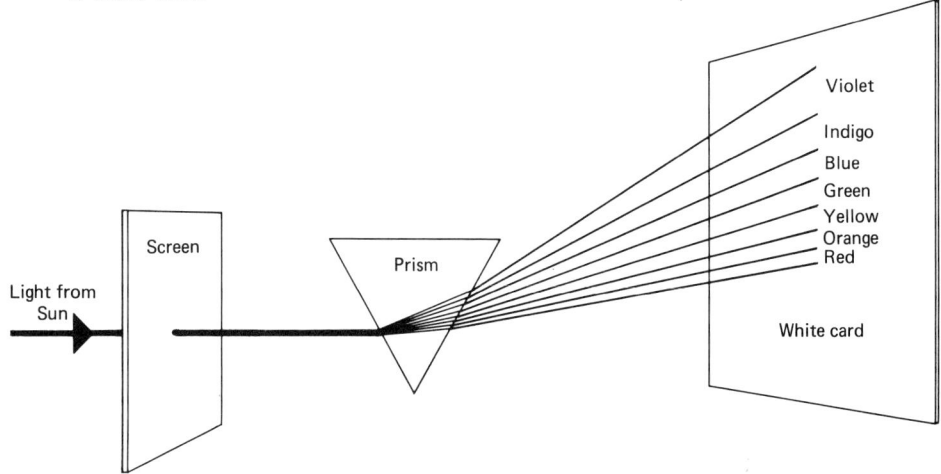

You must not suppose the light of the sun actually "ends" at the red and violet. It is not the light of the sun that ends but the color sensitivity of our eyes. By exposing suitable photographic emulsions we find that the light of the sun continues considerably beyond the visible red into the invisible infrared spectrum region; and far beyond the visible violet into the invisible ultraviolet.

Light, whether visible or invisible, is a form of radiant energy. Many of the properties of radiant energy can be explained on the assumption that it is transmitted across space as a form of wave motion, or vibration, like the waves running along a rope when you move one end rapidly up and down. "But out in empty space what is there to vibrate?" you object. We don't know. All we know is that such an assumption gives a satisfactory agreement between theory and observation.

The waves running along a piece of rope can be described by their wavelength. The length of a wave is the distance between successive high or low places in the waves. The vibrations in radiant energy can also be described by their wavelength. It might seem hopeless to try to measure the length of light rays without some very special instruments. Yet Newton and others succeeded in doing it by performing the simplest experiments, like blowing soap bubbles, for instance. Newton's neighbors, watching him blowing soap bubbles in his backyard, must have thought he had gone completely out of his head. From some experiments with the thin films formed between glass plates, Newton found the wavelength of yellow light is 0.0000225 inches.

Radiant energy originates from such diverse sources, and is detected and used in such widely different ways, that it is hard to believe they are all simply different manifestations of the same kind of energy. Thus we use X rays to photograph your lungs and heart, light rays to examine these photographs, and radio waves to obtain news reports. Yet these three forms of rays differ only in the length of their waves. Scientists obtain information from radiant energy having a range of more than a trillion in wavelength, from the shortest cosmic rays to the longest radio waves. In order of increasing

The Stars and Serendipity

wavelength we have cosmic rays measured in trillionths of an inch, through the gamma rays of nuclear reactions, X rays, ultraviolet rays, visible light, photographic infrared rays, to radio waves measured in miles. There is no sharp dividing line between these rays to which we give various names, for example, between X rays and ultraviolet rays or visible red light and the photographic infrared.

You may be interested in the wavelengths of the different colors which our eyes perceive, as shown below.

Approximate Wavelength of Different Colors

Color	*Wavelength*
Violet	0.000016 inches
Blue	.000018
Green	.000021
Yellow	.000023
Orange	.000025
Red	.000026
Deep red	0.000032

Our eyes are most sensitive to yellow-green light which, by accident or not, is the kind of radiation the sun emits most abundantly. Stars hotter than the sun emit more blue and violet light. Stars cooler than the sun emit a larger percentage of orange and red light.

If we admit sunlight into our spectroscope through a *very* narrow slit, only a few thousandths of an inch wide, we notice a change in the appearance of the spectrum. It is no longer a continuous strip of rainbow colors. Instead in many places it is crossed by narrow, dark lines, as if someone had snipped out narrow sections with a pair of scissors (Fig. 33).

What makes these dark lines?

They are places in the spectrum where light of that particular color is weaker than the average. The light at these places has been partially removed by the atoms and molecules in the sun's atmosphere. Although the lines look black, actually there is considerable light in them. Some twenty thousand lines of different intensities have been recorded in the spectra of the sun and stars.

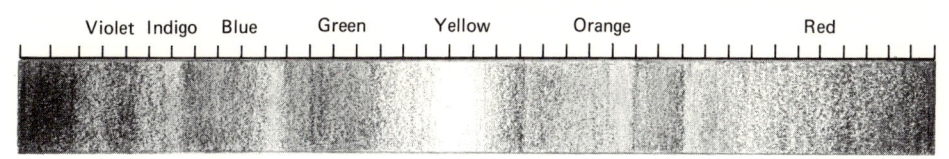

Fig. 32. When white light from the sun is passed through wide slit or hole we get a continuous series of rainbow colors.

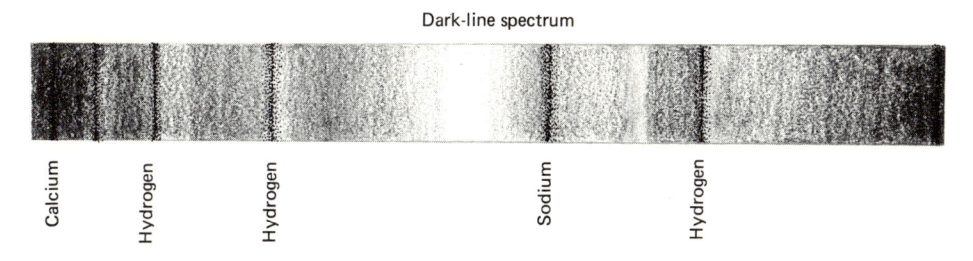

Fig. 33. When white light from the sun is passed through a very narrow slit into the prism, dark lines appear in solar spectrum. Only a few of the most prominent dark lines due to absorption by atoms of hydrogen, sodium, and calcium are shown. Actually there are thousands of dark lines in the sun's spectrum that have been identified with atoms of 66 chemical elements.

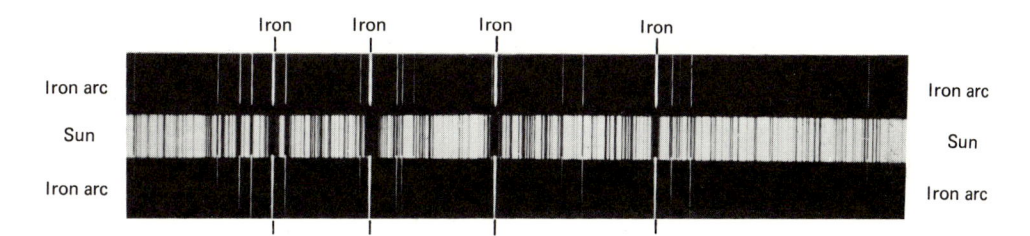

Fig. 34. White lines on dark background arise from glowing iron vapor in laboratory arc lamp. Dark lines in central strip are in the spectrum of the sun. Notice close coincidence of lines in the sun and iron arc.

Investigation shows that most of these lines can be identified with lines known to be produced by such common elements as iron, sodium, calcium, hydrogen, etc., in laboratory spectra. The correspondence in position between lines in the spectrum of iron, for example, and many lines in the spectrum of the sun is so close as to leave no doubt of the presence of iron in the solar atmosphere (Fig. 34). This discovery, made by the investigations of the German scientists Kirchhoff and Bunsen from 1859 to 1862, was of the utmost importance. For it was immediately evident that in the spectroscope astronomers had a powerful research tool by which they could determine the chemical composition of a star almost as easily as if it were a rock in their backyard.

(We hope the reader will forgive us when we tell him that astronomers quit using glass prisms in their spectroscopic equipment long ago. Today they use an entirely different device, called a grating, for breaking up light into different colors. A grating is a bright, reflecting glass surface on which several thousand lines to the inch have been ruled with a diamond point. The spectra produced by reflection gratings are photographed or recorded by some electronic scanning device. We started by talking about spectra produced with a prism because most people are familiar with the prism. But even many scientists have never had any experience with grating spectra.)

Approaching or Receding?

We said that many of the lines in the spectra of the sun and stars correspond in position with lines of common elements produced in laboratory spectra. But they don't correspond *exactly*. There is always a slight systematic difference in position. This difference, small as it is, is of extreme importance. That there should be such a difference was pointed out in 1842 by the Austrian physicist Christian Doppler. If a source of light is approaching an observer, the wavelengths of its rays are crowded together or shortened, resulting in a slight shift of the spectrum lines toward the violet. If the source of light is receding from the observer, the waves are stretched out, resulting in a slight shift of the spectrum lines toward the red (Fig. 35). These slight shifts in the positions of spectrum lines due to velocity are referred to as the "Doppler effect."

The Doppler effect is so sensitive that it enables astronomers to measure velocities of approach or recession of only a few tenths of a mile per second. As we shall see later, there are distant universes

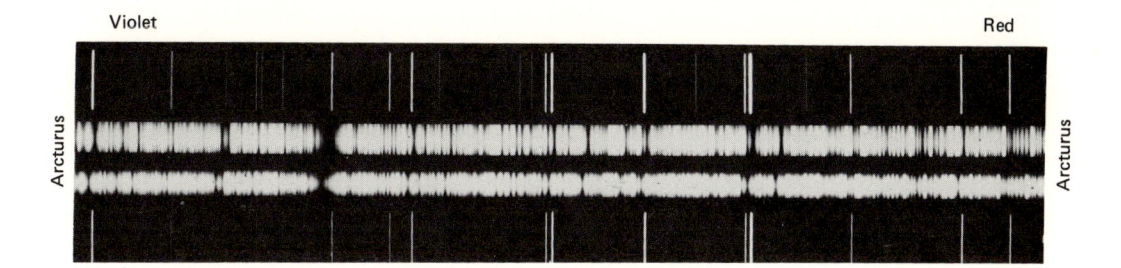

Fig. 35. Bright lines on outer black background, top and bottom, are in laboratory spectrum. Top light strip shows spectrum of bright orange star Arcturus when receding at 11 miles/sec. on July 1, 1939. Bottom light strip shows spectrum of Arcturus when approaching at 20 miles/sec. about six months later. Note shift of Arcturus' lines to red (top), and violet (bottom).

of stars ("galaxies") whose spectra show shifts toward the red corresponding to velocities of recession of tens of thousands of miles per second. Although the spectrum lines show shifts readily apparent to the eye, the change in the color of a source cannot be easily detected.

The late R. W. Wood, noted physicist at Johns Hopkins University, once tried to persuade a judge in traffic court that he went through a red light because his velocity of approach made it appear green. But his defense blew up when the judge discovered that to change the color of a light from red to green would require a velocity of approach of around 147 million miles per hour (corrected for general relativity).

Size of the Companion

By 1900 enough observations had accumulated on the system of Sirius so that astronomers could speak with confidence about the orbit of B, the faint companion, relative to the bright star, A. They knew that it had an uncommonly low luminosity, only about 1/400th that of the sun. Yet its mass was nearly the same as the sun. Its low luminosity they attributed to low temperature. In fact, previous to discovery, Sirius B was regarded more as a huge planet than as a star. By making the companion so cool that it emits scarcely any visible light at all, we can reduce its luminosity to practically any

The Stars and Serendipity

value we please. If the companion is so cool, its color should be deep red.

Is the companion red? That is hard to say.

Why so hard? Can't you tell by looking at it?

Try it yourself. Some clear, dark night examine the stars carefully for color. You will find that only the brightest stars show distinct evidence of color. The others show no definite color at all.

A photograph of the spectrum of Sirius B would tell us its temperature immediately. A hot, white star shows quite a different spectrum from a cool, red star. A hot, white star shows a spectrum that is nearly continuous except for a series of strong, dark lines of hydrogen and perhaps a few weak lines of metals. In a cool, red star there is a multitude of lines due to metals and molecules, whereas the hydrogen lines are weak. But again the proximity of the bright star raises difficulties. The feeble light of B is so overlaid with light from A that instead of getting the spectrum of B we are more likely to get merely another spectrum of A. The situation might be compared to trying to hear a person with a weak voice in a boiler factory. His voice is lost in the background noise.

In 1914 Walter S. Adams at the Mount Wilson Observatory succeeded for the first time in getting a spectrum of B nearly free from the overlying light of the bright star, the separation of the two stars then being a maximum. And the spectrum of B was not in the least what had been expected. It was not like the spectrum of a red star at all. It was the spectrum of a white star, practically as hot and white as Sirius A itself.

This immediately changed all our thinking on Sirius B. For if Sirius B is hot with a temperature of, say, 14,000° F., it must be much smaller than supposed. For a star at 14,000° F. emits thirty-one times as much radiation as a cool star at, say, 5,800° F. But this does not change the facts of observation in the least. The luminosity of Sirius B is *still* 1/400th that of the sun. How are we going to raise its temperature and at the same time hold down its luminosity?

There is only one answer. We must drastically reduce its size. We have to reduce its size until it is only about 32,000 miles in diameter, a body no bigger than the planet Neptune!

Well, so what? Must all stars be of enormous size like the sun?

The trouble arises with the density—the quantity of matter in a given size volume. Suppose we have two pint bottles, one filled with water and the other with the element mercury. The bottle of water would weigh one pound (exclusive of the bottle itself). The bottle of mercury would weigh 13.6 pounds. Volume for volume, therefore, mercury is 13.6 times heavier, or *denser*, than water. If we stuffed the pint bottle full of such flimsy stuff as kapok it would weigh only about 1/10th of a pound. Which tells us at once that the density of kapok relative to water is 0.1. The densest naturally occurring substance is the rare metallic element osmium with a density of 22.56. Ordinarily we don't refer specifically to the density of various substances but simply speak of them as being "light as a feather" or as "heavy as lead," etc.

Now we know the size of Sirius B. And we know it has about the same mass as the sun. So we can calculate its density. Going through the necessary arithmetic we have (hold your breath!),

Average density Sirius B = 30,000 X water

A pint bottle full of Sirius B would weigh fifteen tons. A thimbleful would crush your dining room table. This, you will say, is plainly ridiculous. Something is wrong with our observations or calculations or both. Though estimates of the size and mass of Sirius B have varied slightly, the result is always essentially the same. However you figure it, the density of Sirius B comes out fantastically high.

The White Dwarfs

Stars of such incredible density are called white dwarfs. A score or so are known now. The late Sir Arthur S. Eddington, a leading British astrophysicist, never was altogether happy about the white dwarfs.* Writing in 1925, he remarked that such close packing of

---

* The names *giants* for stars exceeding a certain luminosity and *dwarfs* for fainter stars originated in 1905 with Ejner Hertzsprung, a Danish astronomer.

The Stars and Serendipity

matter as must occur within a white dwarf star is only possible so long as its interior remains at an extremely high temperature. We think of the star as losing heat and gradually cooling down until its interior becomes about the usual density of stellar material. (The average density of the sun is 1.4 times that of water.) For the star to become *less* dense it must *expand*. To expand, it must do work against the force of gravity. Therefore, to cool, the star will need to obtain energy from some source. Where is this energy to come from?

As Sir Arthur said, "We can scarcely credit the star with sufficient foresight to retain more than ninety per cent [of its energy] in reserve for the difficulty awaiting it. It would seem that the star will be in an awkward predicament when its supply of sub-atomic energy ultimately fails. Imagine a body continually losing heat but with insufficient energy to grow cold!"

Degenerate Matter

You would think that, since Sirius B is thousands of times denser than the densest rock, it would be of unimaginable toughness and solidity. You wouldn't be able to scratch it with a diamond drill. Yet the theoretical men insist it is gaseous from center to surface. Now how do they explain a thing like that?

Incidentally, you would have considerable trouble trying to bore a hole in Sirius B or execute any kind of maneuver on its surface. For surface gravity on this body is 25,000 times that on Earth. If you weigh 200 pounds on Earth, on Sirius B you would weigh 2,500—not pounds, but *tons*! An ant or bee would weigh around 50 pounds.

Sirius B is composed of electrons and atomic nuclei, the same elementary particles that make up this book and the air you are breathing. The *material* of Sirius B is basically the same as that on earth. It is the *state* of this material that is so different. Matter in the highly condensed compressed state found in the white dwarfs is called *degenerate*. We never encounter degenerate matter on or in the earth. Neither are we capable of making it artificially in the laboratory. It is doubtful if degenerate matter exists anywhere in the solar system. Speculatively, the pressure might be high enough for its formation in the deep interior of the giant planets.

Perhaps by using a crude mechanical model we can at least make it seem reasonable that matter under extreme conditions may be so highly compressible. Picture the atoms in a gas as men whirling weights attached to long cords. The men would have to be careful to remain sufficiently far apart so as not to get tangled up in one another's weights. Suppose for some reason the men are crammed closer and closer together. Their cords become entangled and broken and the weights torn from their hands. After the men are stripped of their whirling weights they can be packed into a small fraction of the space they occupied before.

The material of Sirius B has been transformed into particles which can be compressed so closely together as to produce almost any degree of density. Yet so tiny are these particles that, even when thousands of times denser than any metal, the material is still "gaseous."

In the superdense material of a white dwarf you would not expect the same laws to hold good that apply to gases in an ordinary state, the air in your tires, for instance. Yet the simple gas laws hold surprisingly well. Much more serious are the deviations from the ordinary laws due to certain restrictions imposed by atomic theory. It is depressing to find that even the supposedly "free" particles in a gas are not entirely free but, like human beings, must still conform to certain rules limiting their behavior.

One of the most remarkable results that has come out of the theory of degenerate matter when applied to white dwarfs is that such stars cannot have a mass exceeding 1.44 times the sun. This curious state of affairs is shown in Figure 36. Notice that for degenerate stars their size *decreases* as their mass *increases*. Thus a star of 0.4 solar masses would have a diameter of 13,000 miles; a star equal in mass to the sun would have a diameter of 7,300 miles; and a star of 1.44 solar masses would be reduced to a point with no diameter at all! Although no such "point" stars have ever been observed, we do know of white dwarfs much smaller than Sirius B, probably no bigger than earth or even the moon.

If there is any truth in our present ideas on stellar evolution, a white dwarf is a star in its senile old age, a feebly luminous body of diminutive size and stupendous density, radiating its energy away so slowly that it may hang on for billions of years before finally reaching total extinction as a "black" dwarf.

The Color of Sirius

A final word about the color of Sirius that does not involve us in complicated theoretical considerations of its structure.

I don't see how you can describe Sirius as any color other than blue-white. Everyone I know is agreed on this. It is true that when low in the sky Sirius flashes other colors—red, yellow, green, blue, etc., but this is purely a prismatic effect due to our atmosphere. Sirius when high in the sky is *white*.

Yet Claudius Ptolemy of Alexandria, one of the greatest astronomers in the world, who flourished from about A.D. 100 to 178, apparently did not see Sirius that way. If you look up Sirius in his star catalogue here is what you will find:

"The brightest and red star in the face called the Dog."

Why did Ptolemy see Sirius red? Evidently Ptolemy was not what opthalmologists call a "red monochromat" whose vision is restricted to a single color, for he doesn't describe other white stars such as Rigel, Vega, and Altair as red. Neither do the astronomers after Ptolemy refer to Sirius as red. Yet Ptolemy was an experienced observer who must have beheld Sirius on thousands of nights.

One conceivable explanation is that stars during a certain period in their evolutionary career swell up and go through a brilliant red-giant stage. Was Sirius perchance a red giant in Ptolemy's time?

You would have a hard time convincing any astronomer that Sirius B could do a changeover from a red giant to a white dwarf

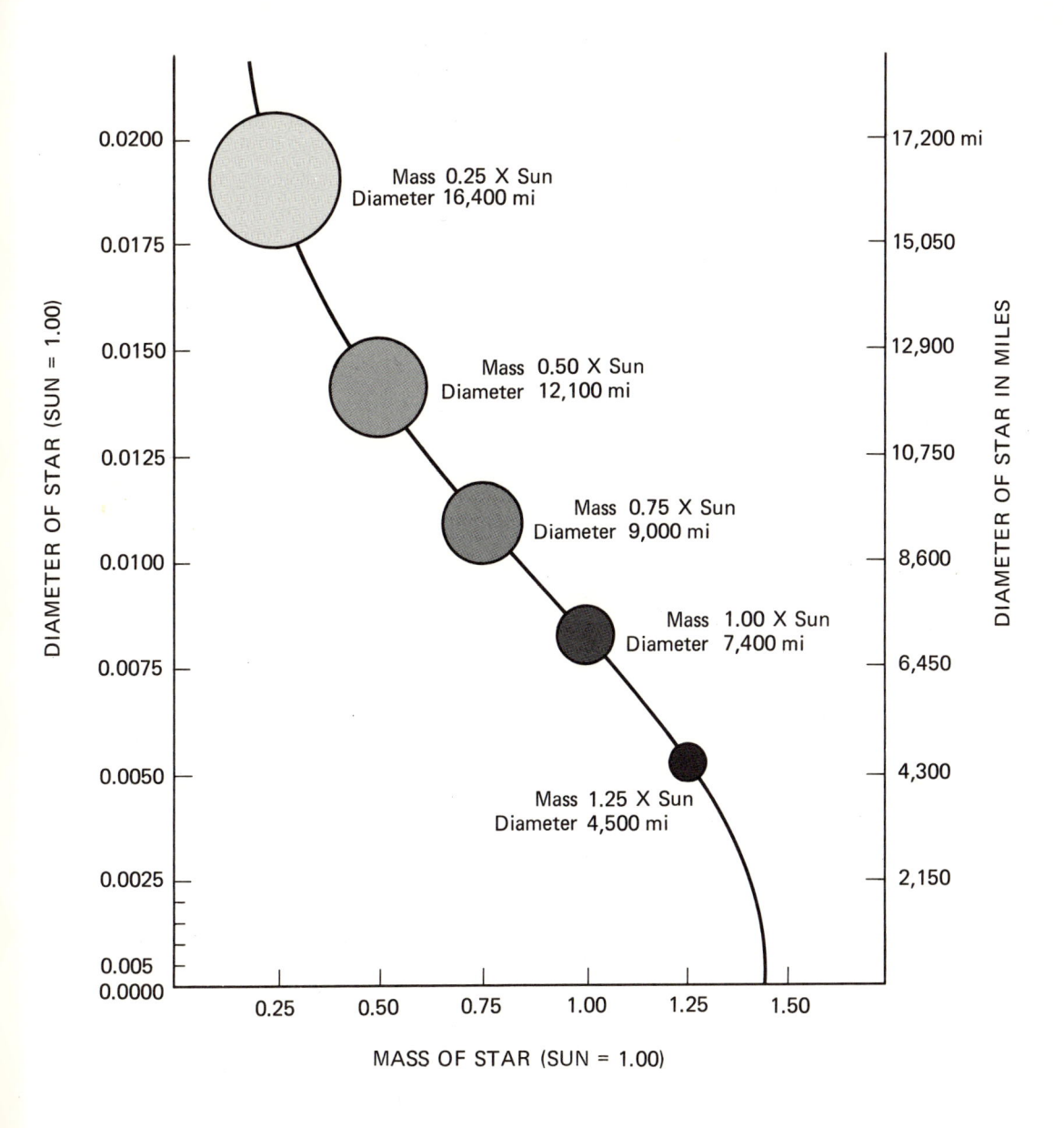

Fig. 36. Theoretical change in size of white dwarf star with increasing mass.

The Stars and Serendipity

in a mere matter of centuries. If the transformation occurred at that rate there should be many other cases on record.

Recently, by chance, I ran across something in this connection that may be worth mentioning. The great Swedish dramatist August Strindberg (1849–1912), in his play *The Ghost Sonata,* Scene 3, has one of the characters say, "But the largest and most beautiful of all the stars in the firmament, the golden-red Sirius, is the narcissus with its gold and red chalice and its six white rays." Apparently Sirius looked the same to Strindberg as it did to Ptolemy. Strindberg was interested in science in a mystic sort of way; for example, he did considerable dabbling in alchemy, astrology, medicine, etc. During a large part of his creative life he suffered from a schizophrenic condition. Yet he had a remarkably retentive memory for certain small facts. Crazy or not, I still find it hard to understand why he called Sirius "golden-red."

Who could have anticipated that out of Bessel's cautious announcement, of more than a century past, about the wavy motion of Sirius, there would eventually come new knowledge of the atom impossible to have been obtained from laboratory investigation? And this may not yet be the end. Three experienced double-star men have reported seeing the companion itself as double. Who can say what surprises may still await us as we probe ever deeper into the Sirius system?

7 · Those Canals of Mars

"Philip Fox told me thirty years ago that he saw the canals at Flagstaff 'stand out like an etching,' and this was the experience of others who worked there."

Edison Pettit, *Publications of the Astronomical Society of the Pacific*, February 1947.

"The network lines, the so-called canals, are not a well-defined class of similar objects, but vary greatly in visibility, width, and definition."

R. J. Trumpler, *ibid.*, 1927

". . . there is no evidence anywhere on the planet of a distinctly geometric pattern; nor is there any evidence of a systematic streakiness in relation to the planetary co-ordinates . . ."

Gerard P. Kuiper, *The Astrophysical Journal*, March 1957

The first *permanent* marking revealed by the telescope on another planet was the dark triangular feature on Mars pointed north that we call the Syrtis Major ("Big Bog"). The dark areas that make up the face of the man in the moon had been familiar from prehistoric times and the cloud belts of Jupiter are easily discerned in a spyglass. But the cloud belts of Jupiter change continually, and technically speaking the moon is not a planet. The Syrtis can be recognized on a drawing of Mars made in 1659 by Christian Huygens, the great Dutch physicist and astronomer. By the end of the seventeenth century several astronomers had recorded Martian features from which they obtained a rotation period for the planet of 24 hours 40 minutes, within 3 minutes of its present accepted value (24 hours 37 minutes 22.6689 seconds).

By the early eighteenth century it was well established that there were permanent dark markings on the surface of Mars, a white spot had been reported at the southern pole, and the planet's rotation period was known with fair accuracy. But with their imperfect

The Stars and Serendipity

telescopes that was about as far as astronomers could go with such a difficult object. More than a century had to pass before telescopes became available comparable in quality with those of today.

Practically all astronomical observations are now made by photography or some sensitive electronic device attached to the telescope. Visual observing is a lost art.

To find great visual observers we must return to the last century. Among the first that comes to mind is the Italian Giovanni Virginio Schiaparelli who will always be known as the discoverer of the canals of Mars. Yet this was not his best piece of work and it is probably safe to say that there must have been times when he regarded his discovery of the canals as more of a liability than a benefit.

The distance of Mars from Earth varies widely, ranging from 35 million miles at closest approach to 62 million miles at its farthest approaches. An approach of Mars, whether favorable or unfavorable, always occurs when Mars is "in opposition." By opposition we mean that Mars is situated opposite in the sky from the direction of the sun. It then appears in the evening as a bright red star rising in the east about the same time the sun is setting in the west. Mars is in opposition about every 780 days. But the close, favorable oppositions occur only at intervals of fifteen or seventeen years, and always in July, August, or September.

The opposition of 1877 was an exceptionally favorable one when the Red Planet was due to come within 35 million miles; in fact, its predicted distance at closest approach on September 5 was 34,800,000 miles. The favorable opportunity for observation due to distance, however, could go for nothing unless the seeing was also favorable. By "seeing" we mean the blurring of the image due to atmospheric disturbances. In "poor seeing" the image of Mars is so badly blurred that only the coarsest markings on the planet can be discerned. In "good seeing" the markings stand out sharp and clear. The seeing is usually better in summer than in winter, and is always very bad in a clear sky after a storm.

Schiaparelli in 1877 was forty-two years old, the Director of

Those Canals of Mars

the Brera Observatory at Milan, a position he had held for fifteen years. Already he was an astronomer of established reputation. Five years earlier the Royal Society had awarded him its gold medal for his researches on the relationship between the orbits of comets and meteoritic streams. He had at his command a refractor of 8.5 inches aperture, a modest instrument, but nevertheless one with which it was possible to do valuable work. In making visual observations of a bright object such as Mars, a huge telescope is not necessarily of such tremendous advantage as people usually suppose; in fact, it has some disadvantages. Owing to its large aperture it is troubled more by atmospheric tremors than are smaller instruments. When observing with a large telescope you can almost always sharpen the image of a planet by diaphragming down, or reducing, the aperture of the lens or mirror.

Schiaparelli seems to have prepared for a heavy observing program like a fighter going into training for a tough bout. He warns against "everything which could effect the nervous system, from narcotics and alcohol, and especially from the abuse of coffee, which I found to be exceedingly prejudicial to the accuracy of observation." (Still sounds like pretty good advice.)

Mars rotates at the rate of about 14 degrees per hour so that the aspect of the disk presented earthward changes rather rapidly. Hence sketching the Martian features is not at all the same as doing a still life of some lemons and dead ducks. Your sketch should not take longer than twenty minutes at the most. So often, however, you get hung up trying to make sure you have represented some detail correctly that half an hour is gone before you know it. But by that time you are not looking at the same disk as when you started.

Figure 37 shows the results of Schiaparelli's observations of Mars at the six oppositions from 1877 to 1888. You must not imagine that he saw all these markings at one time. Rather they are a composite made from probably hundreds of measures and sketches. Neither is the chart supposed to show how Mars looks in a telescope but is intended rather as a map for identification purposes.

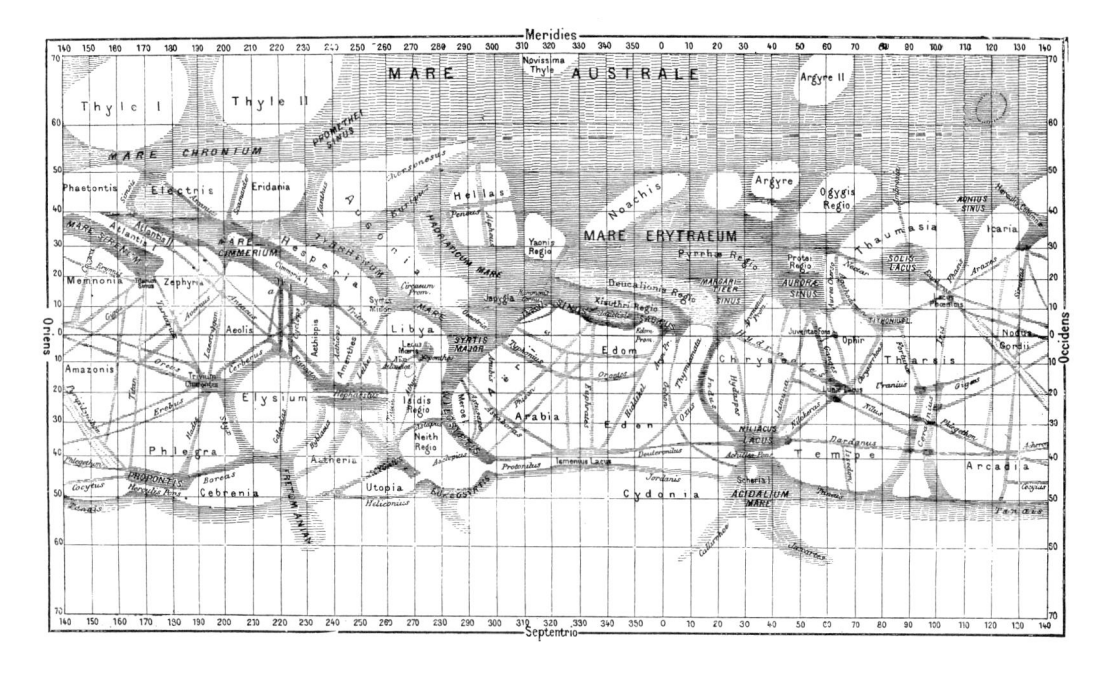

Fig. 37. Schiaparelli's map of Mars. This is a chart intended for identification purposes rather than as a picture of the appearance of the planet.

If you believe the earth is spherical, and everyone else in the world asserts it is flat, it is almost the same as if the earth *is* flat. This was somewhat the same situation in which Schiaparelli found himself. For nine years he was the only astronomer who could see canals. Finally at the opposition of 1886, H. C. Wilson at Cincinnati and J. Perrotin at Vienna, reported sighting canals. Schiaparelli's position in astronomy was a lonely one indeed.

More support was forthcoming at the close opposition of 1892. Observers all over the world began seeing canals. Percival Lowell, at his observatory in Flagstaff, Arizona, seems to have been the first person in this country to see the whole canal pattern simultaneously.

Yet many other experienced observers were unable to glimpse so much as a single canal. They were understandably skeptical of their reality and inclined to believe that the markings originated not on Mars but in the minds of those who so enthusiastically proclaimed their existence.

We wish to emphasize that *canali* was merely the *name* that Schiaparelli applied to the narrow streaks he saw on Mars. He had to call them *something*. If you don't like *canali* go ahead and try to find a better name. Highways? Boulevards? Freeways? Alleys? Not so good, are they? I once devoted several hours to this search and the best I could dig up was "stolons." Stolons is the botanical term for the long, rootlike creepers extending from plants such as strawberries and Bermuda grass.

Schiaparelli himself warned against drawing firm conclusions regarding the true nature of the features he designated *canali*, or "canals," as the word was immediately translated into English. The name was never meant to mean there are real artificial waterways on Mars. The confusion comes in identifying the *name* of the thing with the *thing itself*. The name dandelion means "lion's tooth" (*dent-de-lion*). But because your yard is full of dandelions doesn't mean some lions have been roaming the premises.

Discovery of the canals was serendipity in that it was certainly an unanticipated and happy event. Schiaparelli could not possibly have been looking for such markings on the surface of Mars when he found them. Considering the criticism and abuse he received it may not have been such an altogether happy discovery. Yet, like our other discoveries, it was not in a sense *purely* accidental. He was led to their discovery as the result of a deliberate and carefully planned program of research. Whatever future investigation reveals the canals to be, to Schiaparelli alone belongs the sole credit for being the first to recognize them as a distinctive Martian feature.

But Where Are the Canals?

It would be a waste of time to recount the old arguments pro and con about the canals the planetary men have been hurling back and forth now for nearly a century. We all supposed—hopefully—that once we got photographs of Mars taken from outside the atmosphere the problem would be settled once and for all.

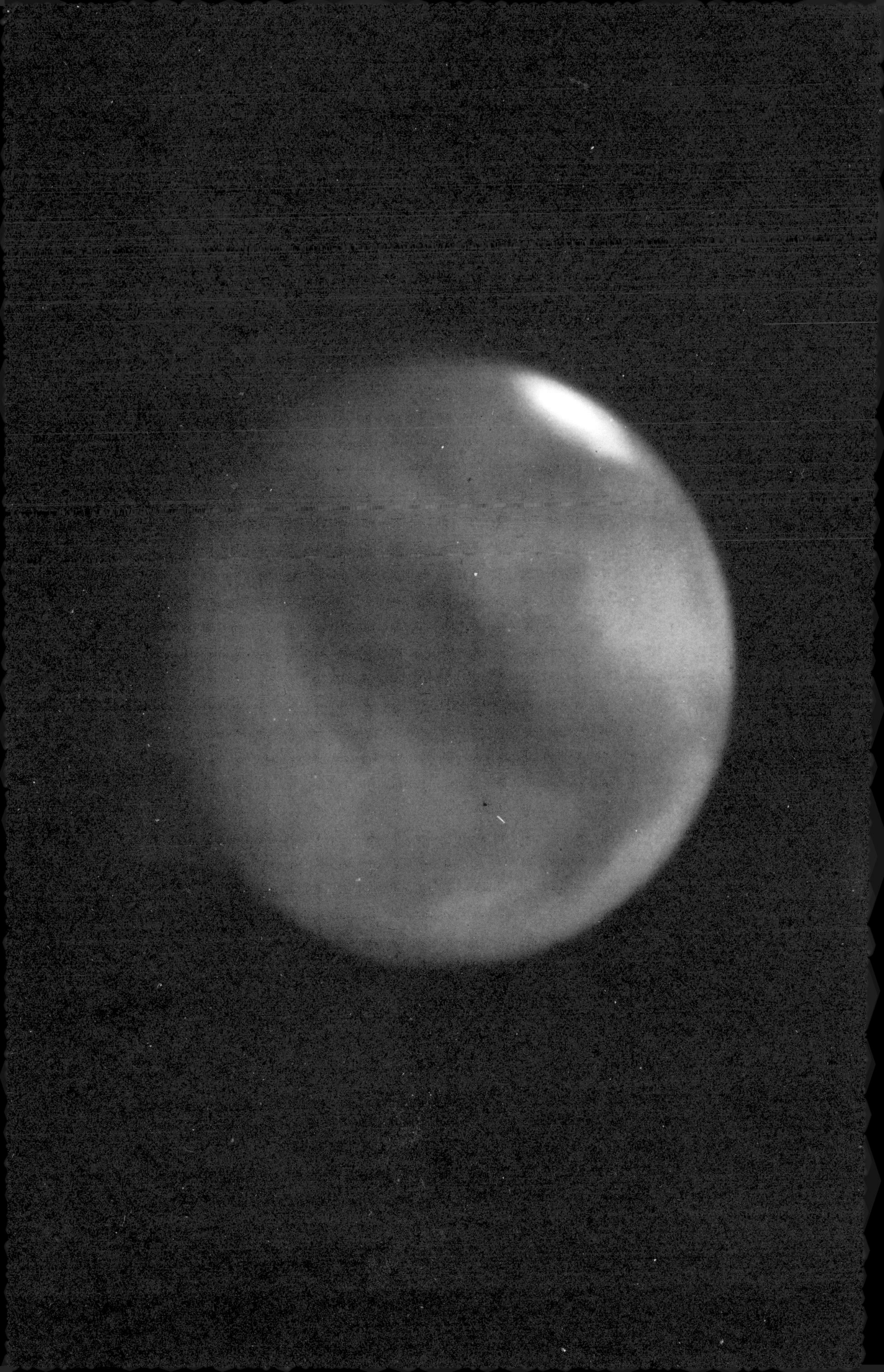

Those Canals of Mars

Well, we have such photographs now (Fig. 38). The beautiful images of Mars obtained in July and August 1969, by Mariners 6 and 7, have revealed a surface bearing a striking resemblance to that of the moon (Fig. 39). On these images there must be hundreds of craters ranging in diameter from a few thousand feet up to more than a hundred miles. Many of the images show incredibly detailed structure. But as for the canals . . . Well, the canals are conspicuous by their absence (Figs. 40a-d).

Fig. 38. Opposite: This photograph of Mars was obtained eighteen days before the opposition of September 1956, using the Mount Wilson 60-inch reflector telescope. The photograph suggests that darker areas are not necessarily "green" as sometimes described, but may be a darker shade of the prevailing yellow-orange. The brilliant white south polar cap is probably a thin layer of frozen water, perhaps hoarfrost.

Fig. 39. Below: Approximate locations of Mariner 6 near-encounter pictures. Heavy lines delineate the wide-angle frames. Small rectangles mark the narrow-angle frames.

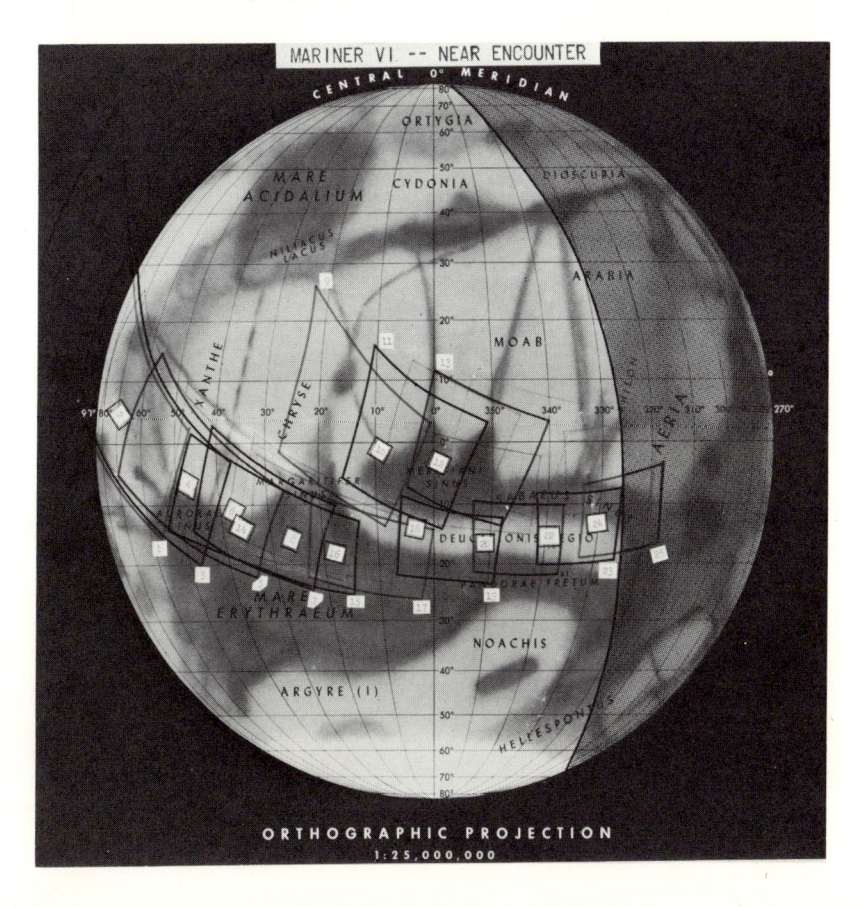

The Stars and Serendipity

Fig. 40a. Right: This Mars mosaic is made up of four overlapping wide-angle pictures taken by Mariner 6 during its close passage of the planet on July 30, 1969. Mariner 6 took the pictures at 84-second intervals as it flew eastward into the nighttime shadow to the right. A grid of black reference dots and trailing white streaks and some electronic noise remain to be removed by further computer processing. Mariner 6 took 74 approach and high-resolution pictures of Mars on July 29–30. Mariner 7 added 126 Mars views on August 2–4.

Fig. 40b. Below: This picture of the Mars south polar cap region, which includes the south pole, shows a wide variety of crater sizes and forms as well as linear and blotchy features not obviously related to cratering. A pair of large craters resembling a footprint in the right center portion of the picture can be seen in more detail in figure 40c. The sunset shadow line (evening terminator) appears on the east (right) of this picture. The grid of small black dots is in the focal plane of the camera, and will be used for making precise measurements on the pictures after further computer processing.

Fig. 40c. Below: The "Giant's Footprint," two adjacent craters foreshortened by oblique viewing of the south polar cap of Mars. The sun is 8° above the local horizon off to the northwest (upper left). This high-resolution picture shows an area approximately 85 x 200 miles.

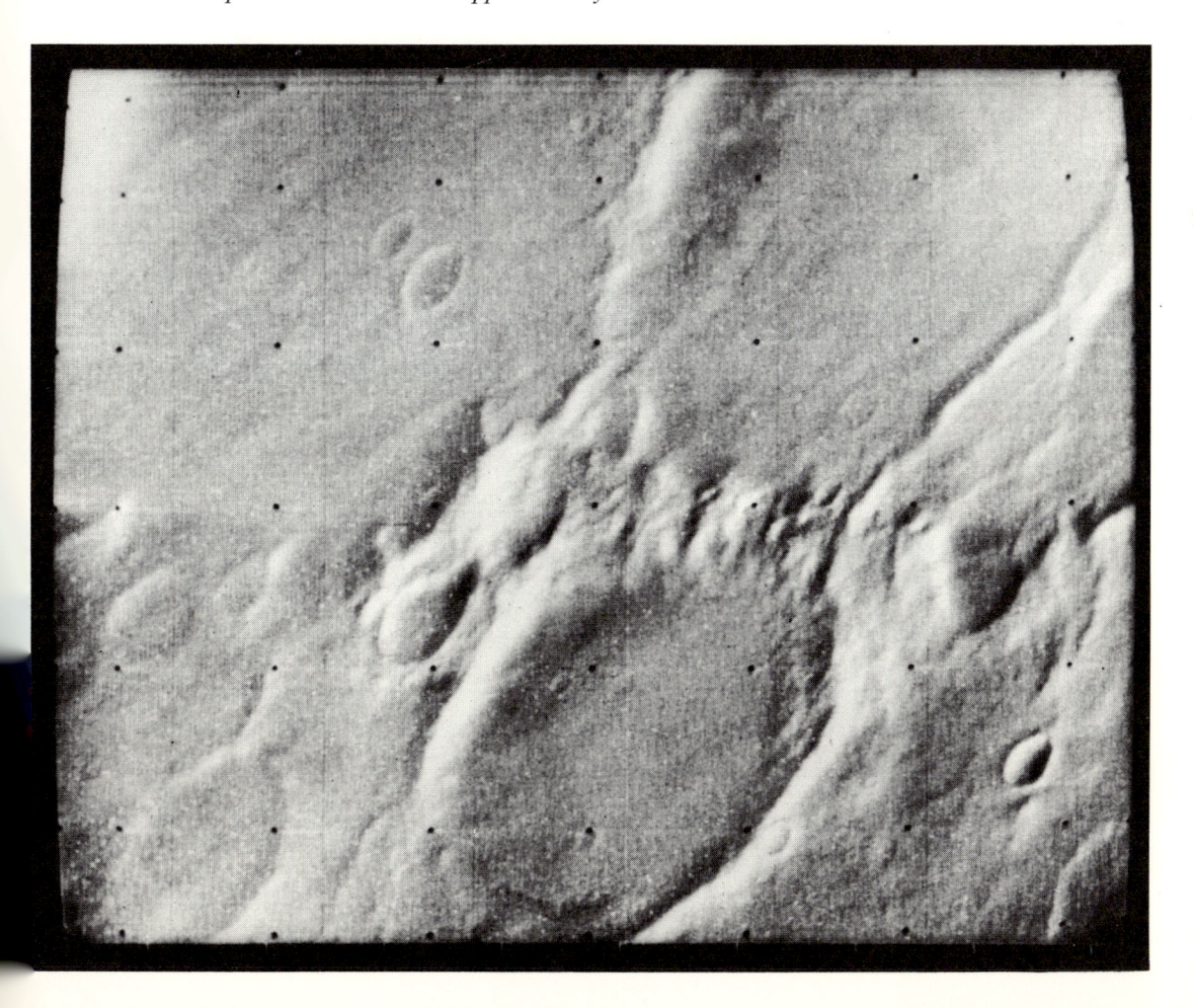

Fig. 40d. Above: This photograph of Mars, made by Mariner 7 in 1969, shows details of some of the markings more clearly than pictures made with ground-based telescopes. The black dot is a blemish in the camera system.

Fig. 41. Below: Mars, photographed by Mariner 7. The four black dots, arranged in a square, are for purposes of measurement.

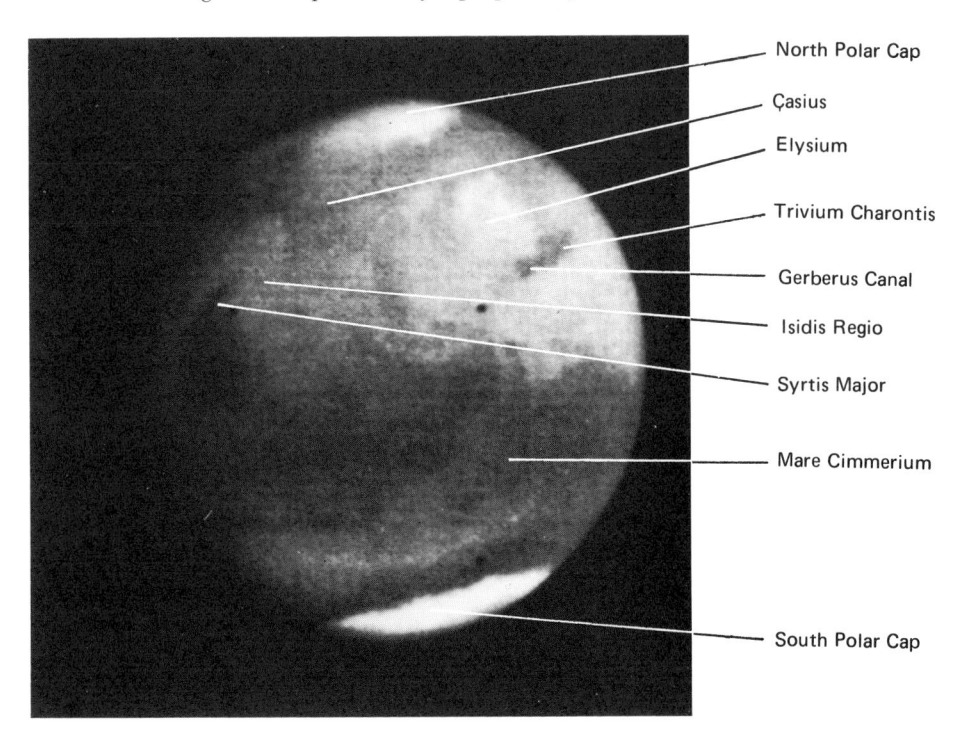

- North Polar Cap
- Casius
- Elysium
- Trivium Charontis
- Gerberus Canal
- Isidis Regio
- Syrtis Major
- Mare Cimmerium
- South Polar Cap

Those Canals of Mars

This writer finds it hard to believe that the sharp line-like markings, which some astronomers have described as standing out on the surface of Mars with the clarity of the rulings on an engraving, are nonexistent. To dismiss them as illusions is too easy. You can explain away anything you like by calling it an illusion. I have never seen the canals sharply defined. I have on occasion seen a few streaks on Mars which if you tried to represent them in a sketch could only be depicted as narrow lines. But I have known astronomers who have seen numerous line-like markings on Mars, and one who had the good fortune to behold the whole canal pattern. They are sane, sober individuals, rather on the conservative side when it comes to observations.

If there are canals on Mars then why didn't the images from Mariners 6 and 7 show them?

I don't know. I thought the most canal-like markings appeared on some of the early Mariner 6 images transmitted to Earth from a distance of 600,000 miles from the planet. These "canals" on close inspection were seen to be due to differences in shading between light and dark areas. But it would be hard to explain in this way the more than 400 canals recorded at the Lowell Observatory.

It would seem to me that the close-up views of Mars would be less likely to show canals than those taken at distances of a 100,000 miles or more. Percival Lowell said the canals, when well seen, are very narrow, only about fifteen to twenty miles in width at the most. On the images of Mars taken at 2,000 miles, the surface area shown is about fifty miles square. Thus a canal, if present, would fill about a third of the screen. Under such high magnification the edge of a canal would be hard to discern unless it stood out in extremely high contrast with the background.

Figure 41 shows an image of Mars taken by Mariner 7 at a distance from the planet of 535,650 statute miles. The research staff at the Jet Propulsion Laboratory refers to the dark irregular marking in the light area at the right of the image as the Cerberus canal. This does not look like any canal to me. To me it looks like the "islet" in the great Elysium bright area called the Trivium Charontis. If it is the Cerberus, I should say it is badly in need of repair.

8 · The Solar Magnetic Cycle

"With the arrival of the high-latitude spots of the next cycle (1913–23), we were surprised to find the polarities with the groups arranged oppositively to those of the spots of the earlier cycle . . . This remarkable result led us to expect some systematic error . . ."

George E. Hale and Seth B. Nicholson, 1938

So often advancement in a field can be traced back directly to the enthusiasm, insight, and courage of just a few individuals. The advance of solar spectroscopy in the last half of the nineteenth century was due largely to such lone pioneers as Lockyer and Huggins in England, Jules Janssen in France, and Father Secchi in Italy. The United States had George Ellery Hale (Fig. 42).

Hale graduated from the Massachusetts Institute of Technology in 1890 at the age of twenty-two, and after some research and teaching, became director of the Yerkes Observatory of the University of Chicago. Lots of men are possessed with boundless enthusiasm for some subject. Unlike Hale, however, few are possessed with the independent means for carrying out their ideas. Hale also possessed a remarkable ability for persuading wealthy people to put money into his projects. Perhaps because he had been born into a wealthy family he was not in the least awed by men of great wealth and felt no embarrassment when it came to asking them for donations.

Hale was convinced that a rich harvest lay open to anyone who applied the powerful new tools of spectroscopy and photography to the study of the sun. After some preliminary studies, Hale decided to establish a solar observatory on Mount Wilson, a prominence of 5,700 feet in the San Gabriel mountains of southern California, overlooking Pasadena and Los Angeles. The site proved so satis-

Sunspot

Sunspot umbra

Spectrum line of spot

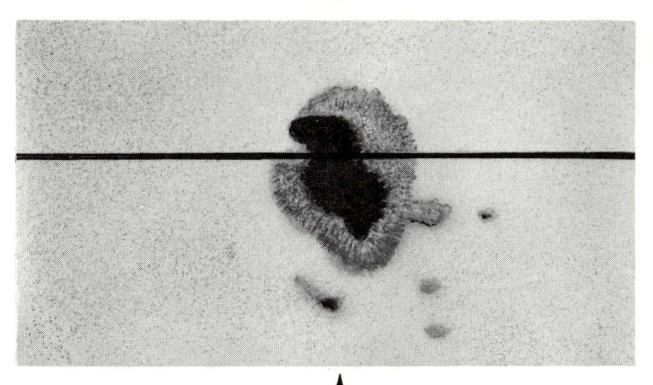

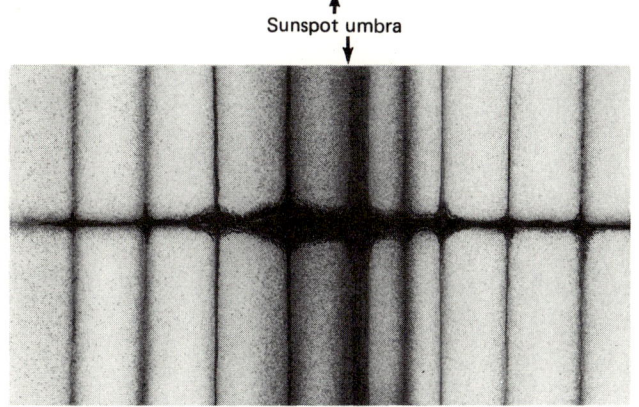

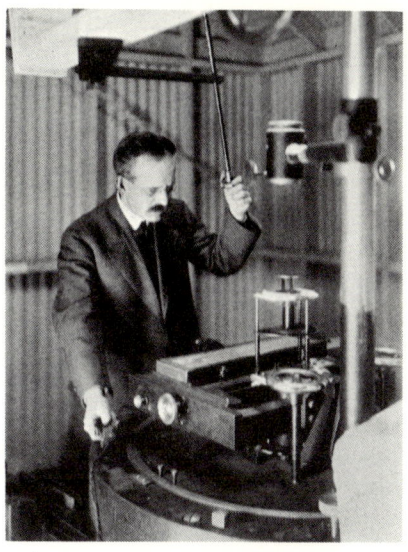

Fig. 42. Above: George Ellery Hale operating the spectrograph with which magnetic fields in sunspots were first measured.

Fig. 43. Left: Notice broadening of spectrum line where it crosses umbra of spot due to magnetic field.

factory that in a few years more powerful solar instruments were installed, as well as 60-inch and 100-inch reflecting telescopes for stellar observations.

Hale's special interest was in the difference in appearance of the spectrum lines in the dark sunspot umbra and in the bright solar surface surrounding the spot (Fig. 43). Doesn't sound very exciting, does it? But it was an effect of intense interest to the astronomers. The spectrum lines of elements such as iron, sodium, manganese, and others are almost always stronger in the umbra than in the bright disk. But the spectrum lines of a few other elements, such as hydrogen, are weaker in the umbra than the disk.

Most such differences in line strength could be readily explained as a temperature effect. The temperature of the bright solar disk, or photosphere, is 11,000°F.; the temperature of the sunspot umbra about 8,500°F. The umbra, being cooler than the photosphere, does not radiate as much light and hence appears darker. The umbra only looks "dark" by contrast with its surroundings. Actually it is brighter than a carbon arc. Laboratory experiments and atomic theory show that spectrum lines at different temperatures should exhibit the same differences in strength that we observe in the umbra and photosphere.

Solar Magnetism

But many of the lines showed other peculiarities that could not be explained by temperature. Most of the lines were wider in the umbra than the disk. This widening depended upon where the spot was located, whether near the edge of the disk or near the center.

Hale believed he had a clue to the spots' behavior in the photographs of the sun taken in the red light of hydrogen gas (Fig. 44). On such photographs the sun looks quite different than on direct snapshots taken in white light (Fig. 45). On the hydrogen photographs the spots often appear surrounded by a vortex structure as if at the center of great solar cyclones. Electrically charged particles in rapid rotation will produce a magnetic field. Are sunspot umbrae the center of powerful magnetic fields?

The field around the poles of a horseshoe magnet are evident from their attraction on particles of iron. But you can hardly use such a method to detect a possible magnetic field in a sunspot 93 million miles away. Fortunately another method of detecting a magnetic field had recently been announced.

In 1896 Pieter Zeeman of Holland had demonstrated that spectrum lines in a luminous vapor are profoundly modified when acted upon by a magnetic field. Ordinarily the spectrum lines appear straight and sharp. But when acted upon by a magnetic field they may be split up into several components, the splitting depending upon the direction as well as the strength of the field. In a weak magnetic field a line may not be split apart into distinct components, but only widened slightly, as shown in Figure 43.

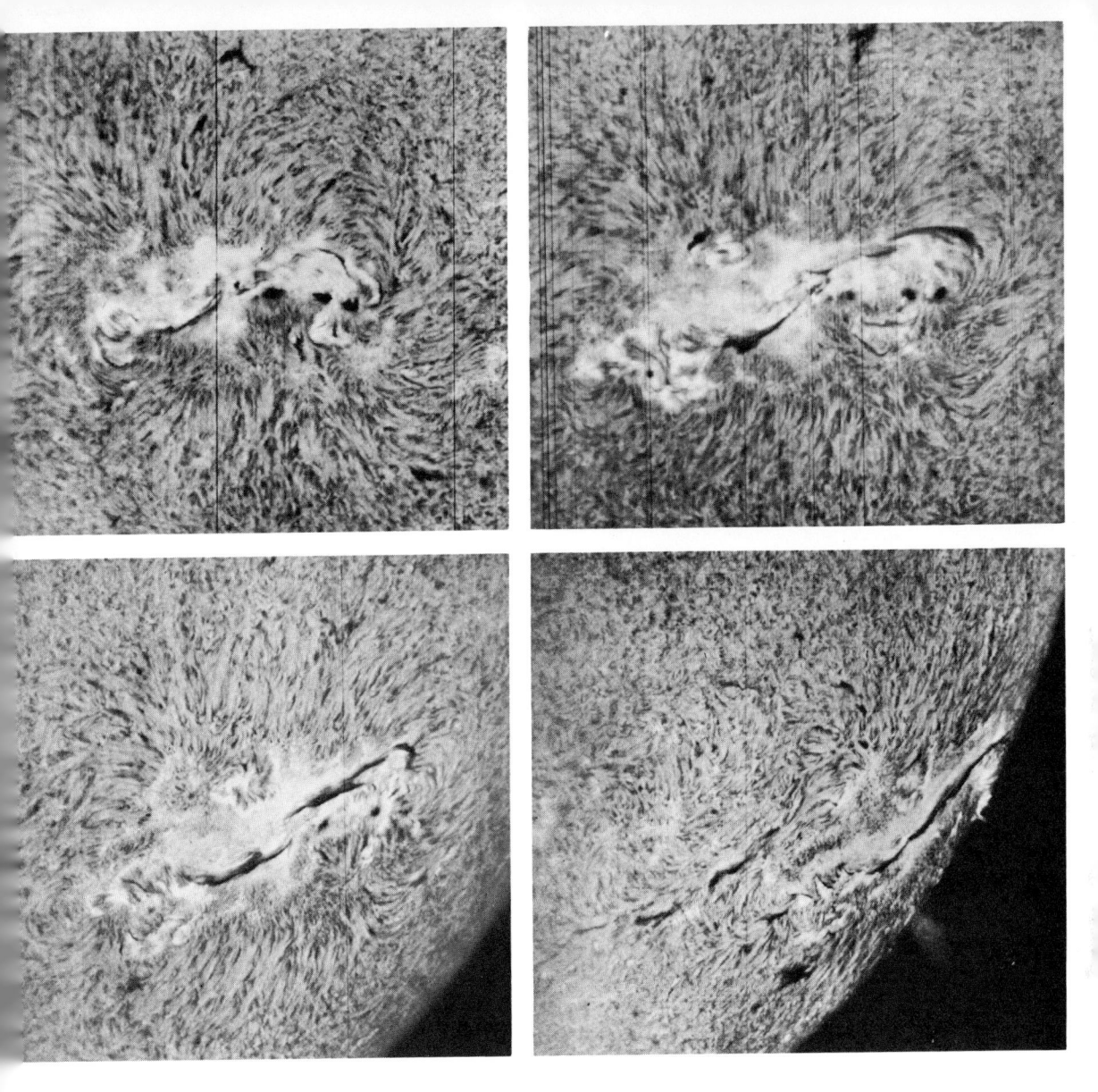

Fig. 44. Above: Photographs of solar surface around spot group showing vortex-like structure, taken in red light of hydrogen atoms only.

Fig. 45. Below: Photograph of spot groups taken in white light, or mixture of all colors.

The Stars and Serendipity

Hale found that lines in the spectrum of the umbra conformed in every respect to their behavior as predicted by the Zeeman effect. After many tests, he sent an account of his results to Zeeman in Holland, asking his opinion. Zeeman was most enthusiastic. "I can say that I have come to the conclusion that Professor Hale has given what appears to be decisive evidence that sunspots have strong magnetic fields . . . ", he wrote to the scientific journal *Nature*.

The origin of the widening of lines in the sunspot spectra was solved. A beautiful example of agreement between theory and observation.

And now . . . what of serendipity?

Spots are not scattered at random over the solar disk. Instead they almost always occur in the form of clusters arranged in two fairly distinct groups (Fig. 46). As these groups are carried across the disk by the solar rotation, it is convenient to refer to the advance portion as the "leading spot," and the rear portion as the "following spot."

Fig. 46. Large sunspot group showing bipolar nature of group.

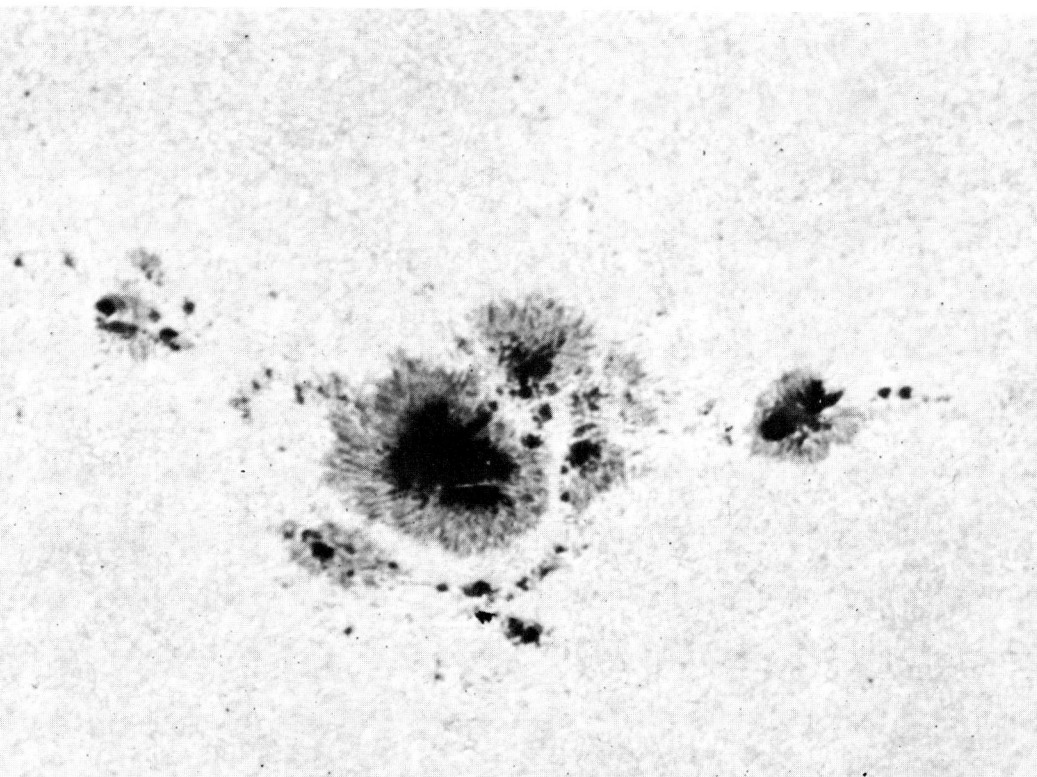

The Solar Magnetic Cycle

Hale and his associates found that the leading and following spots in a group have opposite magnetic polarities, like the north and south poles of a horseshoe magnet. Moreover, the polarities of spot groups in opposite hemispheres of the sun are opposite. When the first magnetic observations on spot groups were made around 1905 in the northern hemisphere of the sun, the leading spots were of South polarity and the following spots of North polarity; that is, the arrangement of the spot groups was (S–N). In the southern hemisphere of the sun the arrangement of polarities was just the reverse: the leading spot was of North polarity and the following spot of South polarity, or (N–S).

Imagine the astonishment of the Mount Wilson observers when the first spot groups of the new cycle began to appear in 1912 (Fig. 47). Their polarities were arranged in the reverse order from those of the old cycle. That is, the new groups in the northern hemisphere had (N–S) polarities, and those in the southern hemisphere were arranged (S–N).

Fig. 47. Magnetic polarities of spot groups in old cycle spot groups from 1901–13. At the end of an 11-year cycle old spot groups occur mostly near solar equator. New cycle spot groups of 1913–23 in higher latitudes showed unanticipated reversal in magnetic polarity.

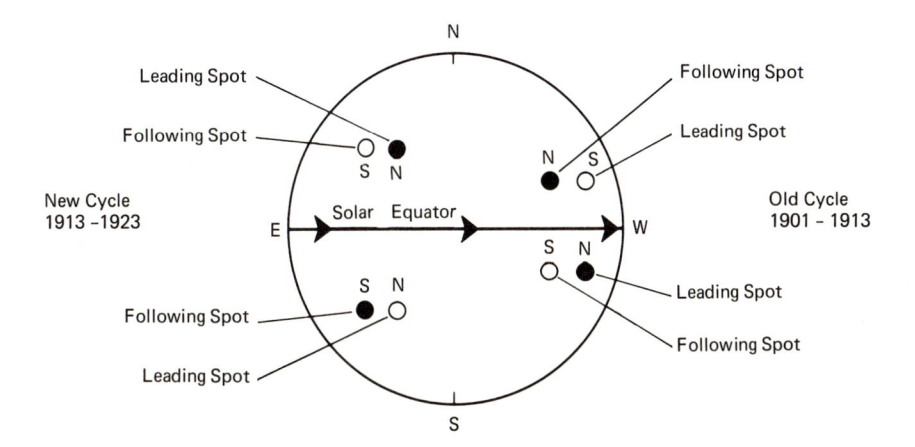

This effect was wholly unanticipated and led the astronomers to fear that some systematic error had gotten into their measures. Repeated checks, however, failed to shake their validity. This reversal in polarity has since been observed at the sunspot minima of 1922, 1933, 1944, and 1954, removing all doubt that it is another fundamental characteristic of the mysterious solar cycle discovered by Heinrich Schwabe.

It comes as an anticlimax to relate that we no longer believe sunspots to be the center of vast hydrodynamical solar cyclones, as Hale originally thought. Other considerations, which we are unable to relate here, force us to reject such a hypothesis. The fact that Hale's early theory is now abandoned is of no particular consequence. Practically all theories eventually meet a similar fate. What *is* of consequence is that Hale's pioneer work led to new basic knowledge of the sun, knowledge that never would have been obtained otherwise.

9 · The Shift Is to the Red

"In the present paragraph I shall conduct the reader over the road that I have myself travelled, rather a rough and winding road, because otherwise I cannot hope that he will take much interest in the result at the end of the journey."

     A. Einstein, *Cosmological Considerations on the General Theory of Relativity*, 1917

It was an unseasonably warm evening in the autumn of 25,000 B.C.

One of our primitive ancestors was lying outside his cave gazing at a sky as yet unstained by smog. He had spent a hard day painting pictures of wounded bison to pep up the oncoming food foray. But somehow he had failed to inject the hunting magic that he desired into his images, and the effort had left him unstrung and depressed. He was tired of doing the same dead and dying animals over and over again anyhow. Cave art needed something new. Like the human figure . . .

Gradually he fell to musing on the stars, tracing out the figures among them that had been familiar to him since boyhood. What was the meaning of all those stars? They must mean something. His artist's eye noticed they were all bright points. Why *points?* Why not bright patches like the glow from a distant fire? Why not—?

There *was* a faint patch among the star points. Gone now. No, there it was again. A hazy patch about the size of the full moon. Showed best out of the corner of his eye. Bit of cloud probably.

But the next night the faint patch was still there at the same place among the stars. He pointed it out to some of his friends but they weren't much interested. What good was it for killing bison or lighting a fire?

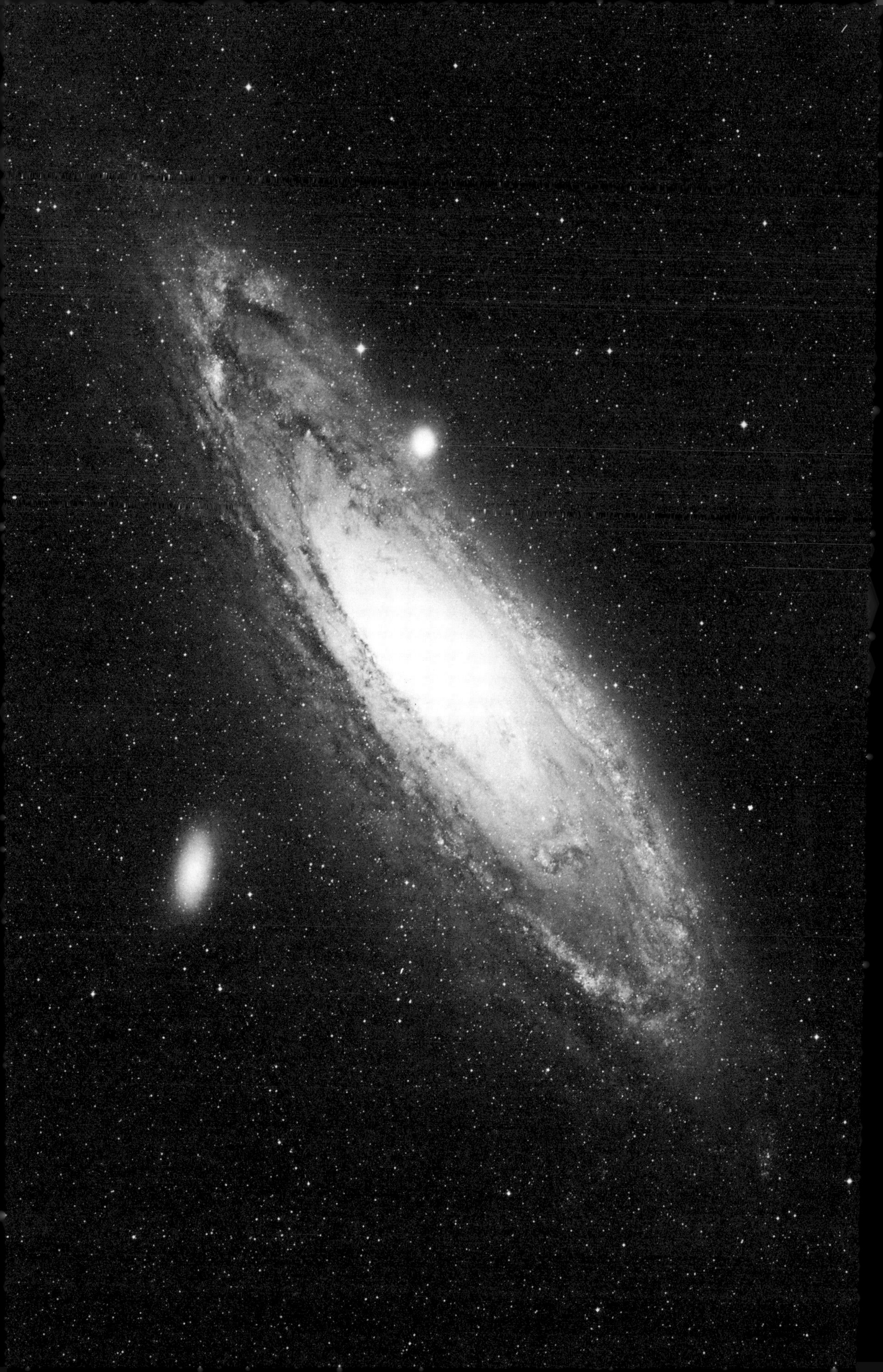

The Shift Is to the Red

The foregoing is an attempt to describe the discovery of the Great Galaxy in the constellation of Andromeda. Whether it was discovered that way we don't know. We shall never know. But it *could* have happened that way. People keep discovering this object in just about that way every year. Under favorable atmospheric conditions it is not particularly difficult to find with your unaided eye. It is easy to pick up with opera glasses or with a telescope under low magnification. High magnification spreads out the faint glow until it becomes so thin you can't tell where the light of the object ends and the sky background begins.

The large reflecting telescopes of the last century resolved many such hazy patches into giant globular star clusters. But others persistently defied resolution even with the most powerful instruments. The spectroscope revealed the light of some of these objects to be concentrated in a few bright lines, proving them to be glowing clouds of gas. But other clouds, or nebulae, such as the one in Andromeda, gave simply a continuous spectrum of the kind emitted by an incandescent solid or liquid. Spectrograms obtained by photography in 1909, however, showed the supposedly continuous spectrum crossed by dark lines such as would be emitted by the combined light of vast aggregations of stars, too distant to show as individual points. Photographs showed many to have a distinct spiral structure, that reminds us of the fiery arms of a pinwheel.

Astronomers held sharply opposing opinions regarding the nature of the spiral nebulae. One group maintained they were "island universes" situated far outside the confines of our Milky Way system. The other group held that they were within the Milky Way. The opposing groups had an opportunity to air their views in the great debate of April 26, 1920. Such debates can clarify an issue; otherwise they are completely useless. You can't prove a truth of nature by arguing about it. The only kind of proof that counts is experiment and observation.

Fig. 48. The Great Galaxy in Andromeda as photographed with 48-inch Schmidt telescope. Note dark lanes and arms in this spiral galaxy distant 2,200,000 light years.

The Stars and Serendipity

The hassle over the distance of the spirals and other nebulae belongs to the history of astronomy now. Suffice it to say that proof of their distance far outside our Milky Way is overwhelming. Our best present value for the distance of the Andromeda galaxy is 2,200,000 light years (12,930,000,000,000,000,000 miles. Henceforth we shall conserve space and effort by expressing distances in terms of light years, 1 LY = 5,880,000,000,000 miles.)* The distances of certain clusters of galaxies are estimated at around 3 billion light years.

First Systematic Attack

In 1912, Vesto Melvin Slipher of the Lowell Observatory, Flagstaff, Arizona, began the first systematic photography of the spectra of the spiral nebulae. (They were called "nebulae" then; the term "galaxy" did not come into general usage until about the middle 1930s.) It is comparatively easy to photograph the spectrum of a star since all its light is concentrated into a point. But trying to photograph the spectrum of a broad, diffuse source is much more difficult.

"The one obstacle in the way of success of this undertaking is the faintness of these nebulae," Slipher said, writing in 1913. "The extreme feebleness of their dispersed light is difficult to realize by one not experienced in such observing, and it no doubt appears strange that the magnificent Andromeda Spiral, which under a transparent sky is so evident to the naked eye, should be so faint photographically. The contest is with the low intrinsic brightness of the nebular surface, a condition which no choice of telescope can relieve."

We have said that the spectroscope—or better, spectro*graph*—furnishes us with two valuable sources of information about a celestial body impossible to obtain in any other way. They are so important as to be worth repeating:

---

*The term light year is rather misleading. Notice that it is a unit of *distance*, the distance that light moving at 186,300 miles per second travels in a year. It is *not* a unit of time.

1. The spectrum lines enable us to identify the chemical elements in the source.
2. The position of these spectrum lines relative to their "rest" positions in the laboratory tell us the velocity of the object toward or away from us. If the lines are shifted to the violet side of their rest positions the object is approaching. If the shift is toward the red the object is receding. (Remember: *red* goes with *receding*.)

Notice that these Doppler shifts, already described, tell us only the radial, or line-of-sight, velocity of an object. They give us no direct information on whatever sidewise motion it may have.

For his first object, as you might expect, Slipher selected the nebula in Andromeda. Perhaps you might like to see the results of his first four plates of 1912. All are velocities of *approach*.

September 17	176 miles per second
November 15–16	184 " " "
December 3–4	191 " " "
December 29–30–31	187 " " "

Notice that some of the exposures were so long they extended over two and three nights. Notice also that he spent New Year's Eve on the job.

Stars commonly show radial velocities of around 20 miles per second or so. But this was the first time that any celestial object with a radial velocity approaching 200 miles per second was known. As Slipher remarked, ". . . the magnitude of this velocity, which is the greatest hitherto observed, raises the question whether the velocity-like displacements might be due to some other cause, but I believe we have at present no other interpretation for it. Hence we may conclude that the Andromeda Nebula is approaching the solar system with a velocity of about 300 km per sec [186 miles per second]." In more familiar terms this is 671,000 miles per hour.

Slipher continued his observations until by 1925 he had secured spectra of forty-one nebulae. Their average velocity was much higher than that of Andromeda—1,400,000 miles per hour *away* from the solar system. The highest velocity recorded was for a nebula known by its catalogue number of NGC 584 with a red shift corresponding to 4,030,000 miles per hour.

The Stars and Serendipity

The remarkable fact was that practically all the nebulae showed velocities of recession, or red shifts. Furthermore, the fainter and presumably the more distant the nebulae the greater were their red shifts. Later, when the Sun's motion within the Milky Way became known and could be allowed for, it turned out that *all* the nebulae were receding. Thus the motion of Andromeda when corrected for the solar motion was changed from a velocity of approach of 186 miles per second to 27 miles per second of recession.

Probing Deeper

By 1925, Slipher had pushed his equipment at Lowell to its limit. Edwin P. Hubble at Mount Wilson, realizing the importance of Slipher's work (V. M. Slipher died in 1969 at the age of 94 while this book was in progress), resolved to continue his program with the 100-inch telescope. By 1929 Hubble was able to announce a relationship between the velocities and distances of the extra-galactic nebulae or *galaxies*. This relationship, like most of the basic laws of physics, is an exceedingly simple one. The velocity is directly proportional to the distance. (The relationship is deceptively simple. It is not so simple when written out in full.) This means that a galaxy 100 million light years away has twice the velocity of a galaxy 50 million light years away. Similarly, a galaxy 200 million LYs distant has twice the velocity of recession of a galaxy 100 million LYs distant, etc.

Hubble's original velocity-distance relationship, which he regarded as a mere "preliminary reconnaissance," has undergone considerable revision through the years. According to some results announced by A. R. Sandage in 1970, which may be changed by the time this appears in print, $H_0 = v/r$, ranges in value from 9.3 miles per second to 24.8 miles per second, per million light years, where v is in miles per second, r in millions of light years, and H_0 is the "instantaneous present epoch of the Hubble 'constant.'"

The velocities and corresponding distances of some galactic

Fig. 49a. Cluster of galaxies in Hercules.

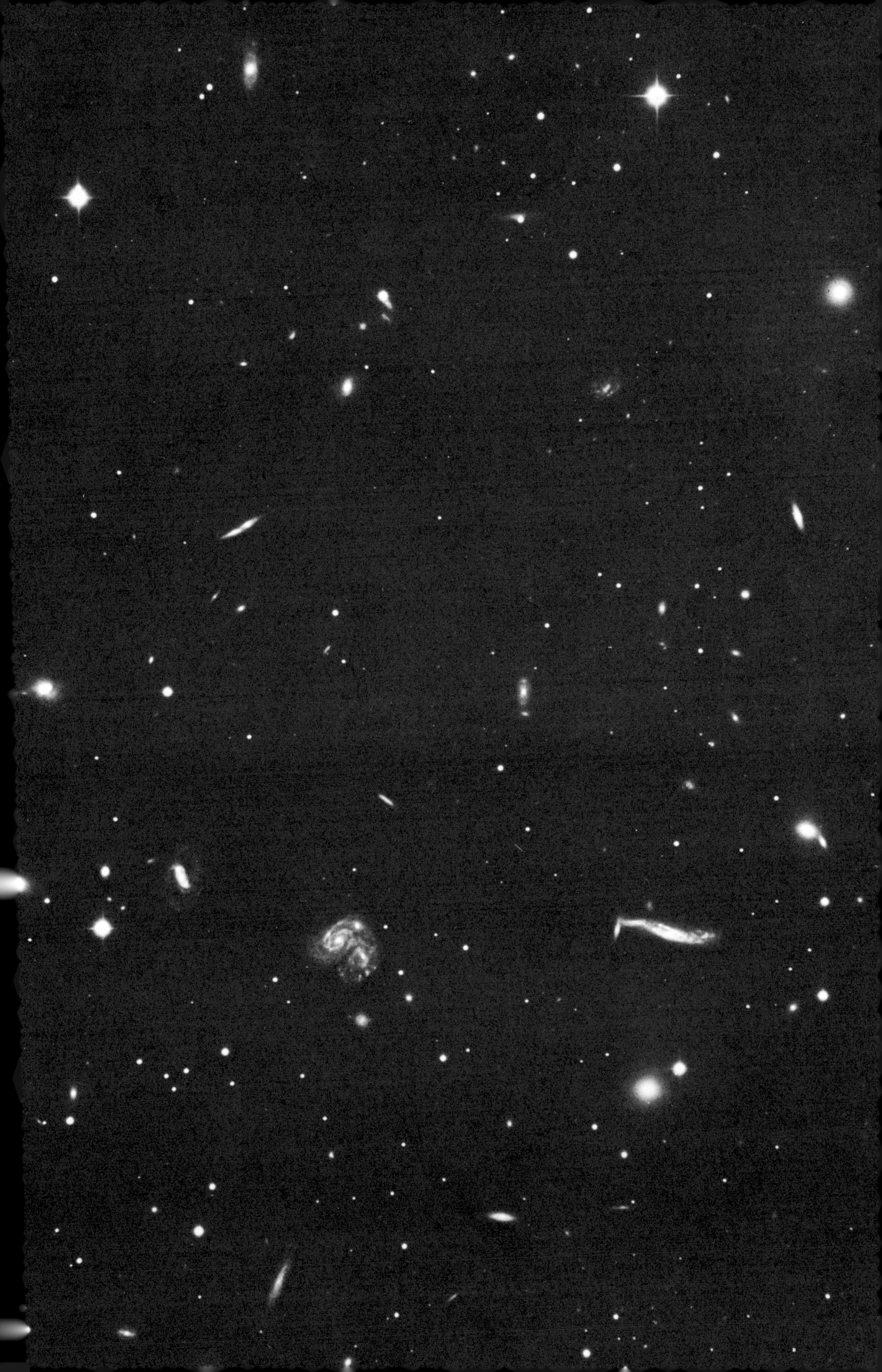

clusters corresponding to a mean value for H₀ of 17.1 are given below.

Galactic cluster	Velocity of recession (miles/sec)	Distance (light years)
Perseus	3,400	200,000,000
Hercules	6,500	380,000,000
Ursa Major No. 1	9,300	540,000,000
Corona Borealis	13,100	770,000,000
Boötes	24,200	1,400,000,000
Hydra	37,900	2,200,000,000

Wherever we look into space we find the galaxies flying away from us, faster and faster, the greater their distance. So far as *appearances* go, at least, our observations indicate we live in an expanding universe (the "Hubble Bubble").

From Hubble's constant, H_0, we can derive a value for the age of the universe. Suppose at some remote time the universe was condensed into a very small volume. Owing to some catastrophe of unspecified nature this highly condensed mass exploded, sending galaxies flying into space in all directions. If some particular galaxy which we now find at a distance, r, received the velocity v, then to travel its present distance must have taken the time

$$T_0 = r/v = 1/H_0.$$

Knowing H_0, we can immediately derive the time T_0 since this catastrophe happened. Thus using the proper units we get a range in T_0 of from 20 to 7.5 billion years for the "age of the universe," whatever that may mean.

The age of the universe! How impressive it sounds. Think how far we have come since Vesto Melvin Slipher first photographed the spectrum of the Andromeda galaxy on that Tuesday evening, September 17, 1912. How could he possibly have foreseen that the little sliver of dark silver grains on his plate was ultimately going to lead us to the time since creation? Not all cosmologists would agree on this rather naive interpretation of the significance of Hubble's constant.

Fig. 49b.   *Cluster of galaxies in Hydra.*

10 · Galactic Radio Waves

"In the opinion of the writer, Jansky's work was a superb piece of observational astronomy which should rank with the discovery of Neptune and the measurement of the parallax of 61 Cygni."

Grote Reber, Leaflet No. 259 of
the Astronomical Society of the Pacific, November 1950

In a few decades after World War II radio telescopes revealed celestial objects and resulted in discoveries wholly beyond the range of ordinary optical telescopes. How were astronomers led to the discovery of radio waves from outer space?

To be brutally frank about it astronomers had nothing whatever to do with the discovery of radio waves from outer space. Instead they practically had to have their existence forced upon them. Cosmic radio waves were discovered by an electrical engineer who had no primary interest in astronomy.

Radio waves from the galaxy were first detected and their origin identified by K. G. Jansky, an engineer from the Bell Telephone Laboratories. In 1931 he set up a receiving apparatus for study of the direction of arrival of terrestrial thunderstorm static. He noticed that some small terrestrial atmospherics persisted, however, even when no thunderstorms were in progress. It would have been easy to disregard this residual static, or to shrug it off as some instrumental defect. But Jansky's curiosity was aroused and he determined to run down the origin of the disturbance. He happened to be observing at a time of year when the static was most intense in the east about sunrise and noon, and faded in the west about dusk. From

Galactic Radio Waves

its behavior it was natural to associate the source of static in some way with the sun.

We have said that the orbital motion of the earth is reflected in the apparent eastward motion of the sun. (This apparent yearly eastward solar motion due to the earth's orbital revolution must not be confused with its apparent daily westward motion due to rotation.) The effect of the sun's eastward motion is to make a day by the sun four minutes longer than a day by the stars. You have the impression that the stars are continually advancing westward by four minutes per day, or two hours per month, to meet the sun. Actually it is the other way around: the sun is advancing eastward by four minutes per day to meet the stars.

Jansky found from his later work that the effect steadily progressed with the seasons. That is, as the seasons passed, the times of appearance and disappearance of the radio emission steadily moved forward in time until after a year the disturbance was back in its original position with the sun. He correctly concluded that the source of emission could not be in the sun or solar system, but was coming from the general direction of the irregular band of light encircling the heavens of the Milky Way.

Jansky announced his results in a technical journal published by the Institute for Electrical Engineers, a type of magazine not likely to be perused by many astronomers. In any case, it met with much the same indifference as Heinrich Schwabe's discovery of the sunspot cycle. Apparently the only person who grasped the significance of Jansky's discovery and its enormous possibilities was a young amateur radio enthusiast, Grote Reber, of Wheaton, Illinois (Fig. 50). Somehow he managed to assemble a radio receiver from whatever odds and ends of equipment he could lay his hands on. With this crude instrument he was able to make a contour map of the sky, showing the radiation distributed along the Milky Way, with especially strong emission in the directions of Cygnus, Cassiopeia, Canis Major, and Orion. (In spite of interference from a dentist's office down the street.)

Fig. 50. First radio-type meridian transit for the study of radio waves, built in 1937 by Grote Reber at Wheaton, Illinois.

He applied to several research institutions for money to improve his equipment, but none was able to see any future in a radio telescope. Finally he secured favorable action largely as a matter of chance. The members of an investigating committee happened to arrive on a rainy day. To their astonishment they found he could get observations right through a sky full of clouds. How wonderful! Application granted.

Before the advent of the radio telescope, the astronomers' situation might be compared to the inhabitants of a "Flatland," whose knowledge of the universe was confined to two dimensions only. They were like bugs crawling around on the surface of a table. They could go north, south, east, or west, but not up or down. Then one day a bug accidentally discovers that he can crawl down the side of the table or climb above it. After some hesitation others follow his lead and the rush is on. Soon they are making all kinds of discoveries, things they never dreamed existed before.

The radio astronomers today are in much the same enviable situation as Galileo with his first telescope. Wherever they look they are almost sure of making an important discovery.

11 · Radio Bursts from Jupiter

"... the probability of these emissions depends strongly on the angular position of Io, the first Galilean satellite of Jupiter ..."
  J. A. Burns, *Science*, March 1, 1968

Early in 1955, B. Burke and K. Franklin were scanning the sky around the Crab Nebula in Taurus, using the Mills Cross antenna of the Carnegie Institution of Washington, D.C. The Crab Nebula, believed to be the remnant of a supernova explosion in A.D. 1054, is a strong source of radio waves. During their sweeps in this region the observers were puzzled by the occasional intense bursts of radiation from some unknown source. Three months of investigation disclosed that this annoying source of disturbance had a motion closely corresponding to that of Jupiter. This wholly unexpected discovery was confirmed from bursts recorded in 1950–51, which had been attributed to static arising in our own atmosphere.

Astronomers soon were busy observing Jupiter with optical telescopes hoping to find some unusual feature on the planet that could be identified with the radio bursts (Fig. 51). (We should be careful not to speak about radio "signals" from Jupiter. It sounds as if there were Jovians with a transmitting set under the cloud layer!) The Great Red Spot was the most likely suspect but repeated observations failed to implicate it with the radio results. Three large white spots in the southern hemisphere likewise fell under suspicion briefly but they also had to be discarded. Various speculative hypotheses were advanced involving Jovian lightning storms and the differential rotation of the planet, but none met with much success.

Recent observations indicate the radiation is closely dependent upon the position of Io, the innermost of the four large moons of

Fig. 51. Jupiter showing parallel cloud belts across disk. Black spot is shadow of one of Jupiter's moons.

Jupiter, revolving 218,000 miles from the planet's surface in a period of 42.5 hours. It appears that Io has magnetic properties which strongly distort Jupiter's magnetic field when in its vicinity. Charged particles from Jupiter's radiation field, trapped in this distorted field, undergo rapid changes in velocity, causing them to emit the observed bursts of radiation in the process.

12 · Fireball Radiation

"The presence of thermal radiation remaining from the fireball is to be expected if we can trace the expansion of the universe back to a time when the temperature was of the order of 10^{10} degrees K." [about 18,000,000,-000°F.]
     R. H. Dicke, P. J. E. Peebles, P. G. Roll, D. T. Wilkinson,
           The Astrophysical Journal, July 1, 1965.

We have seen that observations of the galaxies, initiated by V. M. Slipher in 1912, show them all to be receding from the earth, faster and faster the greater their distance. If we think of this motion as proceeding at a steady rate, it is interesting to run the universe forward or backward to suit our fancy, as if it were motion-picture film. In the future we should expect the universe to be bigger and the matter within it spread farther apart. Conversely, looking backward some 15 billion years, we should expect to find the universe in a highly compressed state. In this remote era it must also have been pretty hot, to put it conservatively, say around 20 billion degrees Fahrenheit. Then there was some catastrophe, or "big bang," and the shattered fragments went flying outward in all directions, producing the expanding universe which we now observe.

 As the universe expanded, the matter within it dispersed into its present highly rarified state. Along with this expansion the universe underwent a drop in temperature to its present very cold condition. There are theoretical reasons for believing, however, that at least *some* of the original primitive fireball radiation is circulating in space. The possible existence of such fossil radiation is of extreme interest to cosmologists. But proof of its presence is hard to establish since radiation from other objects—galaxies, stars, interstellar grains—also contributes to the temperature of space. Nevertheless, some optimists at Princeton began assembling apparatus with which they hoped to detect such radiation.

While the Princeton group was still at work on this project, two observers at the Bell Telephone Laboratories announced they had already done it. They were not looking for fireball radiation. They were not even aware of the Princeton group's intention of conducting such an experiment. Instead they were making measures on the galaxy of an entirely different nature, and the fireball radiation intruded as an unwanted and annoying noise signal. Only they didn't call it by any such exotic name as "fireball" radiation. They referred to it as an "excess antenna noise temperature of 3.5° K."

The unanticipated discovery of this remnant of radiation from the original primitive fireball of billions of years ago has aroused fresh interest in cosmology, the study of the general structure of the universe, a field in which it began to look as if we had about reached the limit of exploration with present observational techniques.

13 · Pulsars

"It is clear that none of the proposed pictures of the pulsar explains well the already known observational data. There is obviously a great opportunity here for further work."

F. D. Drake, *Science*, May 3, 1968

The existence of "pulsars," or rapidly pulsating radio sources, was unpredicted by any theory and their discovery wholly unanticipated. The distinguishing feature about the outbursts from these new objects is the incredible regularity of their occurrence. The first pulsar discovered, known by its catalogue number of CP1919, has a radio pulse repetition rate of 1.33730109 seconds, with a variation of not more than a few parts per million per year. Discovery of this object occurred in the traditional serendipitic manner: as slight deviations from measures made during the course of another investigation.

Since discovery of CP1919 in 1968, several other such radio sources have been detected emitting pulsations at intervals as short as a quarter of a second. The fact that pulsars show no parallax indicates they must lie far outside the solar system. The distance of CP1919 is estimated at 410 light years and another, CP0950, at 100 light years.

The trouble with formulating a theory of pulsars consists in trying to conceive of any *natural* mechanism capable of emitting energy at such astonishingly regular intervals. Can the pulsations then be artificial in origin? Such a notion created considerable excitement for a while and was even jokingly entertained briefly by the Cambridge observers, who, among themselves, referred to pulsars as LGM's, meaning Little Green Men.

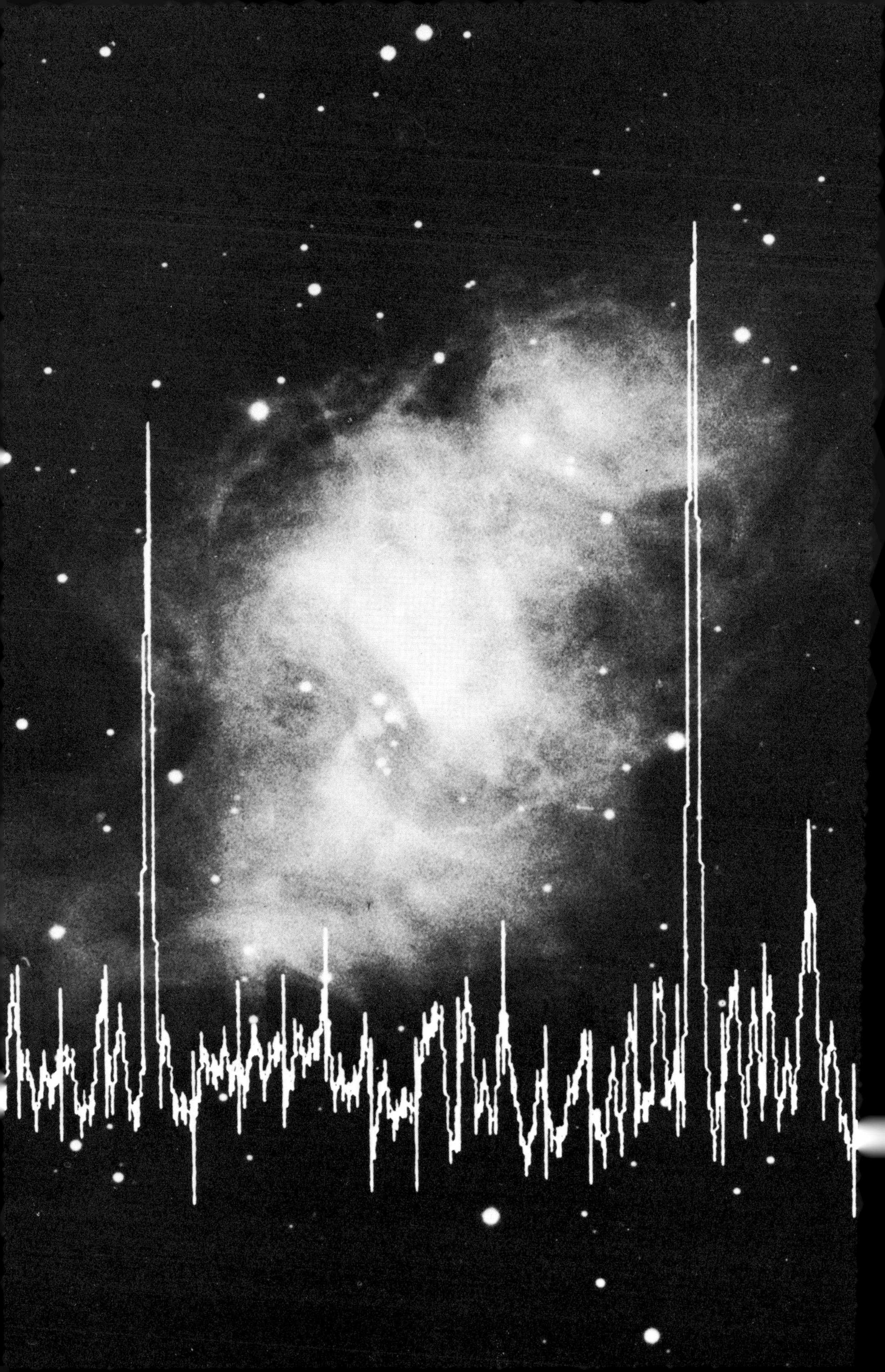

Such a theory, upon close examination, does not stand up any better than the others. Radio astronomers point out that if the inhabitants of some planet with a highly developed technology are indeed trying to signal us, they are going about it in a most peculiar way. Much of the power from pulsars is emitted at wavelengths on which the signals are strongly masked by galactic noise. The transmission occurs over a wide band of wavelengths which is extremely wasteful of energy. And if the pulsars are at the considerable distances estimated for them, the power radiated in strong pulses is enormous, some 10 billion times the total electrical generating capacity of the entire earth. To accept the artificial-source hypothesis we must postulate Little Green Men operating in a bureaucracy that is badly in need of a shakeup in management.

At this writing, the only certainly identified "optical" pulsar is an object bearing the catalogue number of NP 0532, in the Crab Nebula in Taurus (Fig. 52). The Crab Nebula is regarded as the exploded remains of the supernova of A.D. 1054. Present evidence suggests that a pulsar is the magnetic, spinning, condensed relict (remnant or survivor), of a stellar explosion. Future observations will undoubtedly force us to change our current beliefs concerning these fascinating objects, with a good chance of revealing new knowledge at present wholly unknown and unsuspected. Here, however, we are not so much concerned with theories of pulsars as we are with the undeniable fact of their existence.

Fig. 52. The distribution of X-ray pulses from the Crab Nebula in Taurus has the power spectrum shown. This photograph of Crab Nebula is in green light, produced largely by radiation from electrified particles moving in a magnetic field. The pulsar is the lower of the two central stars.

14 · "Mysterium"

"If one civilization wanted to attract the attention of another, which it suspected might be actively engaged in radio astronomy or even actually listening for signals, what better way would there be to attract attention than to violently upset the expected intensity ratios of the four OH lines?"

Alan H. Barrett, *Science*, August 25, 1967

Occasionally we *do* make happy discoveries that *are* anticipated. In 1944 H. C. van de Hulst, a young Dutch astrophysicist, predicted that atoms of neutral hydrogen gas should emit energy corresponding to the radiation of a wavelength of 8.44 inches, sufficiently intense to be observable with radio telescopes. In 1951 the predicted radiation of neutral hydrogen was observed almost simultaneously at three stations. This line proved so valuable in providing information on the distribution of hydrogen in the galaxy that the theoretical men got busy searching for other possible molecules in space.

In 1953 the Soviet astrophysicist S. Shklovskii called attention to the possible existence of lines for radio study due to the hydroxyl (OH) molecule. The OH molecule, being of more complex structure than the hydrogen atom, should produce four lines instead of one. The first search for lines of OH in 1956 was negative, which was not surprising since they were expected to be very weak. The search was also rendered difficult since their positions were known only approximately. Finally, on October 15, 1963, lines attributed to OH were detected in the strong radio source Cassiopeia A.

Where does the serendipity part come in? An effect was predicted. This effect was eventually found. The radio astronomers doubtless were rendered suitably happy over the fact. But you can hardly describe such a discovery as unanticipated.

But it was not quite that simple.

"Mysterium"

The search for lines of the OH molecule in emission was concentrated around a wavelength of 7.198 inches, since this line was supposed to be the strongest of the four. The history of this search is a hodgepodge of false starts, negative results, and misidentification, the reason being that the OH lines displayed such peculiar properties that observers at some stations recorded them without being aware of the fact. Thus when the Berkeley observers turned up an intense line at 7.207 inches, their first thought was that they had discovered some new molecule, which they called "mysterium." Even when the line was finally identified with OH the name still seemed pretty good, in view of its mysterious behavior.

The OH emission source known as W3 has a "brightness" temperature estimated at 20,000,000,000°F. It would be misleading to think of such a "temperature" in the same sense that you think of the temperature of your oven or the water in your swimming pool. Think of it rather as a convenient term for describing the physical properties of W3. Regard its fantastic brightness temperature in somewhat the same way that you regard the "value" of the gold in Fort Knox. When we say that the gold bullion in this underground depository has a monetary value of 10 billion dollars, we are talking about a different kind of money from the money in your pocket.

The anomalous intensities of the four OH lines are as puzzling to astrophysicists as the pulsar bursts. Why do their observed intensities depart so widely from their predicted intensities? It is the sort of problem that only the theoretical physicist can fully appreciate. But the OH lines have their popular side too. Are they being used as a means of interstellar communication? (Readers please note: this idea did not originate with me. It originated with the radio people.)

The argument runs as follows:

The inhabitants of some planet wish to establish communication with others in the Milky Way. Their problem is not so much that of establishing direct contact with higly intelligent creatures such as ourselves, but rather in making other galactic civilizations aware of their existence. Once having attracted somebody's attention they can then proceed with the lengthy and tedious process of developing some sort of intergalactic cross talk. But to attract attention they first have to arouse curiosity.

Suppose you came across the following notice in the *Personal* column of your morning paper:

"Two sisters, Olga and Hilda, siblings, members of scattered family, desire correspondence with marriageable males. Olga, age 12, height 7 feet. Hilda, age 22, height 3 feet. Each has million dollars cash in own right. Address Box W3, Daily Galaxy."

Such an announcement should certainly arouse curiosity. We can hardly imagine any marriageable male simply *ignoring* it. He would ask himself, How could two sisters have such utterly dissimilar physical characteristics? Maybe they are half-sisters. But, no, that can't be because it says they are siblings, the offspring of the same father and mother. But what a pair! Their heights and ages seem to be inverted. Regardless of his situation, what man would want to marry a seven-foot girl of twelve, or a full-grown midget of twenty-two? But that million dollars cash is not to be passed over lightly. He decides to drop Box W3 a line . . .

Analogy with the astrophysical case is obvious.

The inverted age and height of Olga and Hilda attract attention and arouse curiosity. (The anomalous intensities of the OH lines.)

Olga and Hilda have each come into a lot of money. (The strong intensity of the OH emission lines suggests that some amplification, or maser process, is operating, pumping energy into the molecules and upsetting their equilibrium.)

The girls are members of a scattered family. (Although OH emission has been detected in other sources, it is not widely distributed throughout the galaxy, but occurs in small isolated positions.)

Such imaginings seem more like something out of science fiction than serious astrophysical research. Yet the preceding was not abstracted from *Bewildering Stories* or *Esoteric Tales* but from *Science*, the official publication of the American Association for the Advancement of Science, August 25, 1967.*

---

*Alan H. Barrett, *Radio Observations of Interstellar Hydroxyl Radicals*. Dr. Barrett is careful to emphasize the speculative character of his remarks.

"Mysterium"

The science-fiction fan is generally stereotyped as a wild-eyed, slightly cracked individual, capable of believing anything, who writes letters to the editor beginning with "Wow!" On the contrary, speaking from firsthand experience, most science-fiction fans are almost painfully rational personalities with I.Q.s equal to or superior to readers of the *Atlantic Monthly* or *The Kenyon Review*. The one distinguishing characteristic of science-fiction fans is that nothing surprises them.

Here are my own comments on the OH galactic communication idea for what they are worth. I have spent twenty-five years as a professional astronomer on the staff of a large observatory. I have also published several dozen science-fiction yarns under the name of "Philip Latham."

The idea of a vast intergalactic communication network via the OH molecule does not strike me as at all convincing. Magicians say the hardest people to fool are small children and uneducated adults; the easiest are scientists and intelligent adults. I should say that any members of a galactic civilization who think that other civilizations are going to respond to them by screwing up the intensities of the OH doublet spectrum must be optimistically minded indeed. Scientists and engineers can sometimes conjure up the most fantastic hypotheses for a surprising new discovery. Science-fiction fans would be interested but not exactly surprised.

15 · Conclusion

"Common to all these discoveries is that they were made by men doing something and interpretating their results properly. Of course you prefer to plan observations which will certainly be good for something, but it is hard to say if that something will prove to be the most important."

Ejnar Hertzsprung, *Journal* of the Franklin Institute, 1951

We have cited some outstanding examples of serendipity in astronomy. We have seen how serendipity has often radically altered our whole course of thinking in astronomy. Of course, serendipity is not confined to astronomy, or any particular science. There is no reason why it couldn't operate just as well in art, literature, business, or track and field. The dozen or so examples described here are, of course, insufficient for statistical study. Yet it is doubtful if additional data would add significantly to our results.

We have seen how serendipity has operated in the most diverse investigations, from discovery of such minute markings as the canals of Mars to the expansion of the universe. Is there any feature common to these discoveries? What a revolution it would make in our way of life if we could make serendipity work for us on a systematic, rather than a haphazard hit-or-miss basis.

Let us go back now and try to summarize. What can we say about serendipity in general?

1. Serendipity is quite different from having good luck, like finding some money in an old suit of clothes or holding the winning ticket in a raffle. You have to be hard at work looking for something.

2. This "something," whatever its nature, you must consider important. You hope to accomplish it through a carefully planned

Conclusion

research program. You won't get anywhere just sitting around hoping it will happen to you.

3. Fame or a big reputation is no particular help. Most of the discoveries made through serendipity were by comparatively young men who had not as yet achieved any special recognition.

4. The discoveries were all made by men who were alert and intelligent and well-grounded in their field of research.

5. Again and again the discovery was made by running down slight, seemingly inconsequential, deviations from prediction. Not everybody is capable of recognizing minor irregularities of this kind. It requires the type of personality that is always watchful and observant, always "on the ball."

6. After thoroughly checking your observations, so that you are firmly convinced of their reality, don't hesitate to announce your results. Don't try to modify them in an effort to bring them into agreement with accepted theory. Don't cover them up because your boss or some other authority figure won't like them.

7. Maybe we should hedge on No. 1 a bit. A little luck *does* help.

Probably the most essential element in the operation of serendipity is No. 5. Rejoice if your results don't come out just right.

Right now we are at an extremely critical stage in our scientific development where serendipity might suddenly revolutionize our whole way of thinking. This is especially true of extraterrestrial research. Many are asking why we should spend billions in going to the moon, Mars, and other planets. Couldn't this money be spent to much better advantage in improving our own planet?

It is impossible to say what direct practical advantages will come out of extraterrestrial research. If we already knew, there wouldn't be much point in doing it. Nothing of any practical advantage may come out of it. On the other hand, it may lead us to discoveries of such fantastic value we cannot hope to visualize them even in our wildest dreams.

Glossary

Aberration of light An apparent displacement of a star in the direction of the earth's motion, due to combination of velocity of earth and velocity of light.
Angle or angular distance A measure of the difference in direction between two lines that meet in a common point. We use angles to measure the apparent—not actual—size of an object. The moon is 2,160 miles in diameter and the sun 865,000 miles. Both have an apparent angular diameter in the sky of about half a degree (0.5°).
Astrology The pseudo-science that maintains the positions and motions of celestial objects influence life on earth. The predictions of the astrologers do not survive the test of the experimental method.

Black dwarf Believed to be the final stage of a white dwarf star, when it has become extinct and non-luminous.

c The symbol adopted by scientists for the velocity of light in vacuum. c is a constant which has the value of 186,300 miles per second or 299,800 kilometers per second.
Camera obscura A method of forming an image of distant objects by admitting their light through a pinhole or lens into a darkened box.
Celestial Equator Imaginary line in sky directly above earth's equator.
Chimney telescope A telescope formed by setting a lens in a fixed position in a chimney. Limited to observation of stars that pass through observer's zenith.
Constellations Groups of stars to which definite names have been given. Only two or three bear any resemblance to the objects after which they are named. The origin of most constellations is lost in antiquity. Stars in the southern hemisphere, invisible to the ancients, were grouped into constellations in the seventeenth century and many named after "modern" scientific instruments, such as *horologium* the clock, *sextans* the sextant, *antilia* the air pump.

Dark lines in spectrum of the sun or stars. Narrow dark regions in spectra where light has been weakened due to absorption.
Degenerate matter Matter in an exceedingly dense state which, although

gaseous, behaves according to different laws than apply to ordinary gases, such as air.

Density The quantity of matter, or mass, in a given volume of a substance.

Disk As used in astronomy, the apparent flat surface presented by an object such as the sun, moon, or a planet, as seen against the background of the sky.

Doppler effect The apparent change in the wave-length of a spectrum line relative to its rest position, due to motion of the source toward or away from the observer. The line is shifted to the red if object is receding; to the violet if approaching.

Double star or "binary" Two stars revolving around their common center of gravity.

Dwarf As used in astronomy, a star of intrinsically low or moderate luminosity, such as the sun, Procyon, and Alpha Centauri.

Equator Imaginary line drawn around earth (or any planet) midway between poles.

Expanding universe Doppler red shifts of galaxies indicate they are receding faster and faster the greater their distance, as if we lived in an "expanding universe."

Field of view Apparent area of the sky visible in eyepiece of a telescope. The higher the magnifying power of an eyepiece the smaller is the field of view.

Fireball radiation Radiation remaining from supposed catastrophe at birth of expanding universe.

Galaxies Vast aggregations of billions of stars situated far outside our own galaxy or Milky Way system.

Gamma (γ) Draconis A moderately bright star in the northern constellation of the Dragon. It is a good object for observation with a chimney telescope since it passes through the zenith of London.

Giant As used in astronomy, a star of intrinsically high luminosity, such as Aldebaran, Vega, and Rigel (a supergiant).

Globular star clusters Compact aggregations of tens and hundreds of thousands of stars in the form of a fuzzy globe. They occur mostly above and below the plane of our Milky Way system.

Gnomon An upright rod set in the ground from which the altitude of the sun may be measured by length of shadow cast. The gnomon is believed to be the oldest scientific instrument.

Grating (diffraction) As used by astronomers, a bright metallic reflecting surface upon which has been ruled thousands of fine lines. The grating has replaced the prism as a device for spreading light into its different colors for spectrum analysis.

Gravitation   The force of attraction between all bodies. This force increases with the masses of the bodies and decreases as the square of the distance between them.

Gyroscope   A wheel mounted so that it can be set in rotation with its axis oriented in any desired direction. When wheel is rotating the axis maintains its original orientation, even though position of gyroscope mounting is changed.

Horizon   The apparent boundary between earth and sky.

Hubble's constant, H   The factor of proportionality between the distance and apparent recessional velocity of the galaxies, named in honor of Edwin P. Hubble. The estimate of Hubble's "constant" has been changed several times from its original value.

Hydroxyl molecule (OH)   Combination of one atom oxygen joined to one atom hydrogen, detected by radio telescopes in interstellar space.

Infrared rays   Rays of light of longer wavelength than visible red rays.

Interstellar grains   Small solid particles in the space between the stars, probably consisting of frozen nonmetallic substances.

Intra-Mercurial planet   A small planet which astronomers of the nineteenth century supposed was revolving within the orbit of planet Mercury. Best chance of discovering such a body would be in transit across disk of sun, when it would appear as small dark moving spot.

Light year (abbreviated LY)   The distance light travels through space in one year. 1 LY = approximately six trillion (6,000,000,000,000) miles. Notice that the LY is a measure of distance and not of time.

Longitude (celestial)   Angular distance measured from the vernal equinox in eastward direction, parallel to direction of sun's motion.

Magnitude   As used by astronomers, a measure of the apparent brightness of a celestial body. Stars barely visible to the unaided eye are of magnitude 6. Bright stars are of about magnitude 1. A difference of one in magnitude corresponds to a ratio of about 2.5 in brightness.

Mass   For most purposes we can think of the mass of a body as the quantity of matter it contains.

Neutron star   A star so compressed that its atoms have lost their individuality, merging into continuous nuclear matter. A neutron star would be only about six miles in diameter, and of tremendous density, as if entire material of Manhattan Island, rock, buildings, and all, were compressed into the volume of a thimble.

Opposition   A planet is in "opposition" when seen in the sky in the opposite direction from the sun. A planet is in favorable position for observation when in opposition.

Orbit The path described by an object moving around another object under the influence of their gravitational pull.

Parallax Apparent displacement of an object when viewed from different directions. Knowing the parallax of a star is equivalent to knowing its distance.

Penumbra (of sunspot) The border around the central umbra of a spot, lighter than the umbra but darker than surrounding solar surface.

Photographic infrared Infrared rays invisible to the eye but capable of being recorded by photography.

Photosphere The bright visible surface of the sun.

Pluto The outermost planet, discovered in 1930. It is only planet that does not show as a disk in the telescope.

Precession A slow conical motion of the earth's axis of rotation, caused by gravitational attraction of moon and sun on the earth's equatorial bulge.

Pulsars Rapidly pulsating radio sources emitting bursts of energy with remarkable regularity. Their nature is still not understood.

Quasar This word is formed from the phrase "quasi-stellar radio source." Although small in apparent size, the radio emission from quasars corresponds to that of powerful radio galaxies. Quasars show large Doppler red shifts, indicating they are at a great distance in our expanding universe. But the relationship between red shift and distance may be unreliable at great distances. Their nature, like that of other recently discovered objects, is still a matter for speculation.

Radial velocity The velocity of a celestial body in the line of sight. That is, its velocity directly toward or directly away from the observer.

Radiant energy Energy such as light, traveling by wave motion. In a vacuum, it travels with the speed $c = 186,300$ miles per second.

Red shift The shift of lines toward the red (longer wavelength) end of the spectrum in the light from galaxies. Generally ascribed to a Doppler shift due to velocity of recession.

Reflector A telescope in which the rays of light from distant objects are caught and brought together by reflection from the curved surface of a mirror. All the large optical telescopes today are reflectors.

Refractor A telescope in which the rays of light from distant objects are caught and brought together by refraction, or bending, by passing through a curved glass lens. The small telescopes with which most people are familiar are of the lens, or refracting, type.

Rotation Turning on an axis.

"Seeing" The term astronomers use to describe the appearance of star images as affected by atmospheric disturbances. The seeing is "good" when the images appear sharp and steady, "bad" when they appear blurred and shaky.

Sidereal year   The time taken by the sun to move from a certain position relative to the stars to the same position again. The length of the sidereal year is 365.25636 days. This is the actual time required for the earth to revolve around the sun.

Solar motion   Motion of the sun in space relative to the stars in our neighborhood (the "local standard of rest").

Special theory of relativity   The theory advanced by Albert Einstein in 1905, which asserts that measures of time and length depend upon the relative motion of the observer and object observed.

Spectroscope (spectro*graph* when used photographically)   An optical instrument attached to a telescope for analyzing the radiation of celestial bodies. Often described as "splitting light into a series of rainbow colors."

Spheroid   A round body which differs from a sphere in being slightly flattened at the poles, as in the case of the earth (or a grapefruit).

Sunspot cycle   The cycle of about 11 years in the rise and fall of number of spots on the sun.

Sunspots   Dark areas on the surface of the sun lasting from a few days to several months in rare cases. They actually are at a high temperature but look dark by contrast with the hotter, brighter surface of the sun surrounding them.

Tropical year   The "year of the seasons." The time taken by the sun to move from the vernal equinox back to the vernal equinox again. Length of tropical year is 365.24220 days. Owing to precession, the tropical year is 20.4 minutes shorter than the sidereal year.

Ultraviolet rays   Rays of light of shorter wavelength than visible violet light.

Umbra of sunspot   The dark central region of a solar spot.

Vernal equinox   The point, or date, at which the sun crosses the celestial equator from south to north each year, usually on March 21.

White dwarf   A star of about the same mass as the sun and with a surface temperature high enough to appear "white." The diameter of a white dwarf is only about that of a planet. Such stars are of exceedingly high density.

Zenith   Point in the sky directly overhead.

Zodiac   The "zone of animals" nine degrees wide on either side of the sun's path among the stars. Despite its name, four of the twelve signs of the zodiac refer to people—Gemini, Virgo, Aquarius, and Sagittarius—and one to an instrument—Libra the Balance.